Wissenschaftliche Reihe Fahrzeugtechnik Universität Stuttgart

Reihe herausgegeben von

Michael Bargende, Stuttgart, Deutschland

Hans-Christian Reuss, Stuttgart, Deutschland

Jochen Wiedemann, Stuttgart, Deutschland

Das Institut für Fahrzeugtechnik Stuttgart (IFS) an der Universität Stuttgart erforscht, entwickelt, appliziert und erprobt, in enger Zusammenarbeit mit der Industrie, Elemente bzw. Technologien aus dem Bereich moderner Fahrzeugkonzepte. Das Institut gliedert sich in die drei Bereiche Kraftfahrwesen, Fahrzeugantriebe und Kraftfahrzeug-Mechatronik. Aufgabe dieser Bereiche ist die Ausarbeitung des Themengebietes im Prüfstandsbetrieb, in Theorie und Simulation. Schwerpunkte des Kraftfahrwesens sind hierbei die Aerodynamik, Akustik (NVH), Fahrdynamik und Fahrermodellierung, Leichtbau, Sicherheit, Kraftübertragung sowie Energie und Thermomanagement – auch in Verbindung mit hybriden und batterieelektrischen Fahrzeugkonzepten. Der Bereich Fahrzeugantriebe widmet sich den Themen Brennverfahrensentwicklung einschließlich Regelungs- und Steuerungskonzeptionen bei zugleich minimierten Emissionen, komplexe Abgasnachbehandlung, Aufladesysteme und -strategien, Hybridsysteme und Betriebsstrategien sowie mechanisch-akustischen Fragestellungen. Themen der Kraftfahrzeug-Mechatronik sind die Antriebsstrangregelung/Hybride, Elektromobilität, Bordnetz und Energiemanagement, Funktions- und Softwareentwicklung sowie Test und Diagnose. Die Erfüllung dieser Aufgaben wird prüfstandsseitig neben vielem anderen unterstützt durch 19 Motorenprüfstände, zwei Rollenprüfstände, einen 1:1-Fahrsimulator, einen Antriebsstrangprüfstand, einen Thermowindkanal sowie einen 1:1-Aeroakustikwindkanal. Die wissenschaftliche Reihe „Fahrzeugtechnik Universität Stuttgart" präsentiert über die am Institut entstandenen Promotionen die hervorragenden Arbeitsergebnisse der Forschungstätigkeiten am IFS.

Reihe herausgegeben von

Prof. Dr.-Ing. Michael Bargende
Lehrstuhl Fahrzeugantriebe
Institut für Fahrzeugtechnik Stuttgart
Universität Stuttgart
Stuttgart, Deutschland

Prof. Dr.-Ing. Hans-Christian Reuss
Lehrstuhl Kraftfahrzeugmechatronik
Institut für Fahrzeugtechnik Stuttgart
Universität Stuttgart
Stuttgart, Deutschland

Prof. Dr.-Ing. Jochen Wiedemann
Lehrstuhl Kraftfahrwesen
Institut für Fahrzeugtechnik Stuttgart
Universität Stuttgart
Stuttgart, Deutschland

Jonas Maier

Optimization of Zonal Electric/Electronic Architectures Using a Coupled Modeling Approach of Package and Wiring Harness

Springer Vieweg

Jonas Maier
Institute of Automotive Engineering
(IFS), Chair in Automotive Mechatronics
University of Stuttgart
Stuttgart, Germany

Zugl.: Dissertation Universität Stuttgart, 2024
D93

ISSN 2567-0042 ISSN 2567-0352 (electronic)
Wissenschaftliche Reihe Fahrzeugtechnik Universität Stuttgart
ISBN 978-3-658-48187-2 ISBN 978-3-658-48188-9 (eBook)
https://doi.org/10.1007/978-3-658-48188-9

Responsible Editor: Friederike Lierheimer
This Springer Vieweg imprint is published by the registered company Springer Fachmedien Wiesbaden GmbH, part of Springer Nature.
The registered company address is: Abraham-Lincoln-Str. 46, 65189 Wiesbaden, Germany

Acknowledgment

This thesis was completed during my work as a research fellow at the Chair of Automotive Mechatronics at the Institute of Automotive Engineering at the University of Stuttgart under the supervision of Prof. Dr.-Ing. H.-C. Reuss. During that time, I had the opportunity to study electrical/electronic architectures in vehicles and their development intensively, which inspired the idea for this thesis.

I am grateful to Prof. Dr.-Ing. H.-C. Reuss for providing me with the opportunity to research this exciting topic at the institute. The freedom I was granted in selecting my research topic allowed me to explore future ideas through this thesis. His continuous support was invaluable, and without it, the successful completion of this thesis would not have been possible.

I would also like to thank Univ.-Prof. Dr.-Ing. M. Kreimeyer for assuming the role of co-supervisor and for his valuable advice in the final phase of this work. Thanks to his support, the structure of this work could be significantly improved, resulting in its current form.

I want to express my gratitude to all the colleagues at the two institutes, IFS and FKFS, and the UNICARagil project at various universities. Together, we were able to overcome many challenges during my time at the institute, and you have contributed significantly to my professional and personal development, for which I am thankful.

I am especially grateful to Daniel Gehringer for his support in the UNICARagil project and for the enriching exchange we have had over the past few years, which had a profound impact on this work and on me. Without him, I would not have been able to successfully complete either of the two projects. I also want to acknowledge Lukas Brandl, whose discussions helped me identify and address many weaknesses in my original ideas.

I would also like to thank Christina Schrey, Dr. Daniel Trost, Dominik Scheuble and Samuel Maier for reviewing this thesis. Furthermore, I want to thank

Benjamin Krüger and William Siegle, who contributed important parts to this thesis through their student work. I additionally wish to thank all other students who have supported me in the UNICARagil project or through their student work.

Lastly, my special thanks go to my friends and family who have not directly contributed to this thesis but have ensured that I find distraction in my free time. Without this distraction, I would not have had the endurance needed to complete this thesis.

Stuttgart Jonas Maier

Contents

List of Figures

List of Tables

List of Abbreviations

ABS	Anti-lock braking system
AD	Automated driving
ADAS	Advanced driver assistance systems
BMS	Battery management system
CAN	Controller Area Network
CPS	Cyber-physical system
DC/DC	Direct current/direct current
DCU	Domain control unit
DIN	German Institute for Standardization Registered Association (German: Deutsches Institut für Normung e.V.)
DSRC	Dedicated short range communication
ECU	Electronic control unit
EEA	Electric/Electronic architecture
EMC	Electromagnetic compatibility
EN	European standard (German: Europäische Norm)
ESP	Electronic stability program
EUR	Euro
EV	Electric vehicle
GNSS	Global navigation satellite system
HMI	Human machine interface
HV	High-voltage
HVB	High-voltage battery
IFS	Institute of Automotive Engineering
ISO	International Organization for Standardization

LIN	Local Interconnect Network
LV	Low-voltage
MLOC	Million lines of code
MOST	Media Oriented Systems Transport
OEM	Original equipment manufacturer
PT	Powertrain
PTC	Positive temperature coefficient
SAE	Society of Automotive Engineers
SC	Supercapacitor
SOF	State of function
SSE	Sum of squared errors
SSSP	Single-source shortest path
SUV	Sport utility vehicle
USD	US Dollar
VDI	Association of German Engineers (germ. *Verein Deutscher Ingenieure*)
WLTP	Worldwide Harmonized Light Vehicles Test Procedure
ZCU	Zone control unit

List of Symbols

	Greek Letters	
σ	Standard deviation	-
μ	Mean	-
α_{veh}	Road slope angle	-
α_i	Individual component complexity	-
$\beta_{i,j}$	Interface complexity factor	-
Δ_{bnd}	Bundling deviation	-
Δ_b	Bending radius approximation deviation	m
Δ_{lb}	Scaled distance for mutation	-
$\|\|\Delta\|\|_{1,max}$	Maximum within cluster Manhattan distance	m
$\|\|\Delta\|\|_1$	Within cluster Manhattan distance	m
ϵ	Neighborhood environment	-
$\eta_{bat,V}$	Package efficiency factor for battery volume	-
$\eta_{bat,m}$	Package efficiency factor for battery mass	-
η_{el}	Electric energy conversion efficiency	-
γ	Tupel of x, y and z coordinates	-
λ	Weights optimization function	-
ρ_{Cu}	Density of air	$\frac{\text{kg}}{\text{m}^3}$
ρ_{Cu}	Electric resistivity copper	$\Omega \cdot \text{m}$

	Indices
axle	Axle
b,el	Electric balance
bat	Battery
bnd	Bundled
c	Cable
cell	Battery cell
cmp	Complexity
crg	Cargo
d	Indice for data cables

d, i, j	Counting indices
e	Spatial length expansion
ee	Elektric/Elektronic architecture
el	Electric
Eth	Ethernet
I	Input
lv	Low voltage components
lvb	Low voltage Battery
max	Maximum
min	Minimum
n, k	Limiting indices
O	Output
p	Indice for power cables
pow	Power
S	Cluster
s	Start
sc	Supercapacitor
sv	Specified value
t	Target
tot	Total
vl	Voltage level
wh	Wiring harness
whl	Axle
z	Zone
zcu	Zone control unit

	Latin Letters	
$A_{cab,max}$	Maximum cross section of all cables	mm^2
$A_{i,j}$	Adjacency matrix	-
A_{veh}	Projected frontal area	m^2
A_{wh}	Angle of a wiring harness	-
A	Cross section	mm^2
B_{wh}	Breakouts of a wiring harness	-
B	Balance	-
$C_{bat,aux}$	Battery auxiliaries cost	EUR

C_n	Nominal battery capacity	Ah
C	Cost	EUR
$D(x)$	Distance of a data point x to its nearest centroid	-
$D_{cab,max}$	Maximum outer diameter of all cables	mm
$\mathcal{D}$	Set of data cables	-
$E(A)$	Sum of the singular values of an Adjacency matrix	-
E	Energy	kWh
$\mathcal{F}$	Aggregated optimization function	-
F_a	Acceleration resistance force	N
F_d	Air resistance force	N
F_g	Gradient resistance force	N
F_res	Resistance force	N
F_r	Rolling resistance force	N
F	Flexibility	-
G_c	Children generation	-
G_p	Parent generation	-
H	Domain of a function	-
$I_{HV,max}$	Maximum current of the high voltage components	A
I_{vc}	DC/DC converter current	A
$\mathcal{O}$	Time complexity (big O notation)	-
P	Electric power	W
$\mathcal{E}$	Set of power cables	-
R^2	Coefficient of determination	-
R_{el}	Electric resistance	Ω
SOF	State of function	-
S	Set of cluster	-
U_{vc}	DC/DC converter voltage	V
$\mathcal{U}$	Set of graph edges	-
V	Volume	m^3
V_{el}	Voltage	V
$\mathcal{V}$	Set of graph vertexes	-
W_{drv}	Mechanical work of vehicle drive	J
$\mathcal{W}$	Set of graph edge weights	-
$\mathcal{X}$	Set of data	-

$\mathcal{Z}$	Set of zone numbers	-
a_{veh}	Vehicle acceleration	$\frac{\mathrm{m}}{\mathrm{s}^2}$
b_f	Bundling factor	-
b_n	Number of cables per edge	-
c_1	Component complexity sum	-
c_2	Pair-wise interaction complexity	-
c_3	Arrangement of interfaces complexity	-
c_r	Rolling resistance coefficient	-
c_w	Drag coefficient	-
c	Complexity	-
d_{grid}	Grid approximation distance	m
e_i	Mass factor of powertrain inertia	-
$f_{bat,C2S}$	Mass factor battery, cell to system level	-
f_c	Cost factor	$\frac{\mathrm{EUR}}{\mathrm{m}}$
f_m	Mass factor	$\frac{\mathrm{kg}}{\mathrm{m}}$
f	Function	-
g_{cf}	Random number from Laplace distribution	-
g_{mf}	Random number from uniform and power distribution	-
g	Gravity of earth	$\frac{m}{s^2}$
$h(\gamma)$	Voxel of coordinate tupel γ	-
k^*	Cluster number for pre-clustering	-
lb	Lower boundary	-
l	Length	m
m_{load}	Mass of additional load	kg
m	Mass	kg
n_{Bat}	Number of battery cells	-
$n_{f,1}$	Number of components used in flexibility analysis step	-
$n_{f,2}$	Number of components correctly assigned in flexibility analysis step	-
n_{veh}	Number per vehicle	-
$n_{\tilde{x}}$	z coordinate	-
$P_{\varnothing}$	Average electric power	W
p	Path	-

$r_{b,dyn}$	Minimum bending radius for dynamic load	m
r	Radius	m
t_u	Usage time	%
ub	Upper boundary	-
u	Graph edges	-
v_{veh}	Vehicle velocity	N
v_c	Relative velocity	$\frac{m}{s}$
v_{wind}	Wind velocity	$\frac{m}{s}$
v	Graph vertexes	-
w	Edge weights	-
x_1	Starting x coordinate	m
x_2	Finishing x coordinate	m
$\tilde{x}$	Data point	-
x	Cartesian x coordinate	m
y_1	Starting y coordinate	m
y_2	Finishing y coordinate	m
y	Cartesian y coordinate	m
z_1	Starting z coordinate	m
z_2	Finishing z coordinate	m
z	Cartesian z coordinate	m

Kurzfassung

Die Komplexität heutiger Fahrzeuge ist stark abhängig von der Perspektive des Betrachters. Der Markt nimmt sie als Maschine wahr, welche im Wesentlichen der Fortbewegung dient und einige weitere Komfortfunktionen bietet. Die technische Entwicklung hingegen sieht in ihnen einen Verbund vieler Teilsysteme mit starken Abhängigkeiten. Dies ist begründet durch die hohe Anzahl von erforderlichern Entwurfs-, Koordinierungs- und Verifizierungsmaßnahmen zur Umsetzung eines Produkts, in dem alle Systeme lückenlos ineinandergreifen. Zur Realisierung, jeder für den Kunden wahrnehmbaren Funktion, müssen eine Vielzahl von Einzelfunktionen entwickelt und kombiniert werden. Verschiedene Disziplinen, wie Elektronik, Mechanik und Software, sind in einem Umfeld von hohem Kosten- und Zeitdruck eng miteinander zu verknüpfen. Dies resultiert in einem Spannungsfeld, in dem es gilt einen technischen und wirtschaftlichen Kompromiss zu finden, welcher den Kunden zufriedenstellt. Ein zentraler Begriff hierbei ist die Integration, also das Einbinden von technologischen Systemen unter Berücksichtigung technologischer Schnittstellen und diverser Interessengruppen. Die Identifikation dieses Kompromisses entspricht einer Suche nach dem Optimum in einem interdisziplinären und multikriteriellen Lösungsraum.

Die Elektrik/Elektronik-Architektur eines Fahrzeugs dient der Realisierung der Daten- und Energieübertragung in Fahrzeugen. Vor einigen Jahrzehnten betraf dies nur wenige Komponenten, heute hingegen über 100 elektronische Steuergeräte sowie etliche Aktuatoren und Sensoren. Jedoch ist nicht nur die reine Anzahl gestiegen, sondern durch neue Funktionen und eine stärkere Vernetzung auch die Komplexität. Komplexität kann als ein Maß für die systemische Schwierigkeit eines Systems und der Wechselwirkungen zwischen den System-Komponenten verstanden werden. Somit steigt diese bei zunehmender Anzahl von Komponenten und Wechselwirkungen. Dieser Begriff darf nicht mit Kompliziertheit verwechselt werden, da dies ein Maß für die Schwierigkeit beim Verstänis eines Systems oder Wirkzusammenhangs ist. Für die Entwicklung bedeutet dies eine Vervielfachung des Aufwands, da die

Komplexität sowohl der Wechselwirkung innerhalb des Systems als auch zu anderen Systemen in den letzten Jahren exponentiell angestiegen ist. Abbildung 1 zeigt die Trends charakteristischer Größen für die Komplexitätsbeurteilung von Elektrik/Elektronik-Architekturen. Es fällt auf, dass die konventionellen Größen wie Stecker, Sicherungen und Niederspannungs- bzw. LV-Leistung ab 1980 kontinuierlich zunehmen. Diese Trends sind auf die fortlaufende Integration weiterer Funktionen zurückzuführen. Die Anzahl der Daten und Signale sowie der Hochvolt-Leistung nimmt im Zeitraum von 2000 bis 2020 um ein Vielfaches zu. Die Verbreitung von elektrisch angetriebenen Fahrzeugen sowie die stärkere Vernetzung und die Einführung assistierender Fahrfunktionen sind die Ursache für diesen Anstieg. Die Prognose für 2030 zeigt vor allem hinsichtlich der Menge von Daten und Software bzw. Million lines of code (MLOC) einen extremen Anstieg, verursacht durch die fortschreitende Digitalisierung und Vernetzung sowie eine stärkere Automatisierung von Fahrfunktionen.

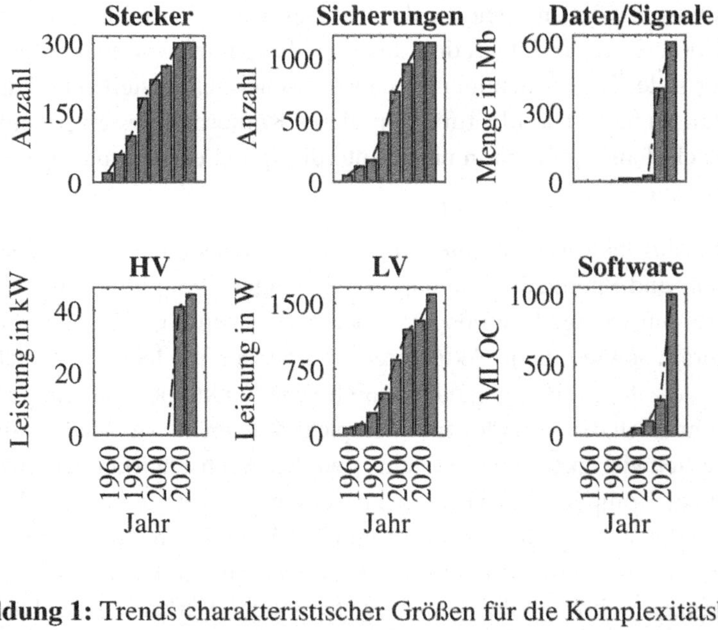

Abbildung 1: Trends charakteristischer Größen für die Komplexitätsbeurteilung von E/E Architekturen, Daten basierend auf [1, 2]

Da die Anzahl der Systemelemente und deren Wechselwirkung kontinuierlich zunimmt, steigt auch die Komplexität bei der Integration von Systemkomponenten fortlaufend an. Aktuelle E/E-Architekturen wurden unter anderen Rahmenbedingungen, also einer geringeren Komponentenanzahl und Integrationstiefe entwickelt, wodurch die Steigerung der Systemkomplexität in einer Erhöhung des Integrationsaufwands resultiert. Zum einen strebt die technische Entwicklung die Konsolidierung mehrerer Funktionen in einzelnen elektronischen Steuergeräten an, um sowohl die Softwareentwicklung als auch die Vernetzungsstruktur zu vereinfachen. Zum anderen hat sich der Kabelstrang in den letzten Jahrzehnten zu einer der aufwändigsten Hardware Komponenten im Fahrzeug entwickelt, sowohl bezogen auf die Entwicklung als auch auf die Produktion. Die Steigerung des Integrationsaufwands betrifft allerdings nicht nur neu hinzukommende, sondern alle Systemkomponenten. Es ist daher davon auszugehen, dass diese Trends auch in Zukunft die Komplexität von Fahrzeugen weiter steigern und aktuelle Architekturkonzepte somit nicht mehr geeignet sind.

Ein Architekturkonzept, welches diese Probleme lösen soll, ist die zonale E/E-Architektur. Der Kern dieser Idee ist es, die Strukturierung der Architektur nicht wie bisher nach technologischen Kriterien, z. B. anhand der Übertragungstechnologie bzw. -protokolls oder anhand funktionaler Kriterien, z. B. Antrieb, Komfort, Infotainment, vorzunehmen. Stattdessen wird das Fahrzeug in Zonen eingeteilt, welche mittels geometrischer Eigenschaften definiert sind. Abbildung 2 zeigt die Prinzipdarstellungen einer Domänen-zentralisierten und einer Zonen-zentralisierten Architektur. Die Domänen-Zentralisierung ermöglichte die Reduktion des Kommunikationsbedarfs, da ein Großteil der Kommunikation innerhalb einer Domäne stattfindet und somit auf diese reduziert wird. Durch die steigende Vernetzung nimmt die domänenübergreifende Kommunikation immer weiter zu, wodurch dieses Konzept an seine Grenzen stößt. Bei der Zonen-Zentralisierung erfolgt ein Großteil der Kommunikation innerhalb einer Zone und die interzonale Kommunikation mittels eines Netzwerks mit hoher Datenrate. Gekoppelt mit einer Service-orientierten Software-Architektur kann die zugehörige Software hardwareunabhängiger entwickelt werden. Ein weiterer Vorteil ist die Komplexitätsreduktion des Kabelstrangs, welche durch die Aufteilung des Kabelstrangs in mehrere Zonen-Kabelstränge ermöglicht wird. Darüberhinaus steigt die Modularität von Hard- und Software, da Komponenten in anderen Zonen wiederverwendet werden können.

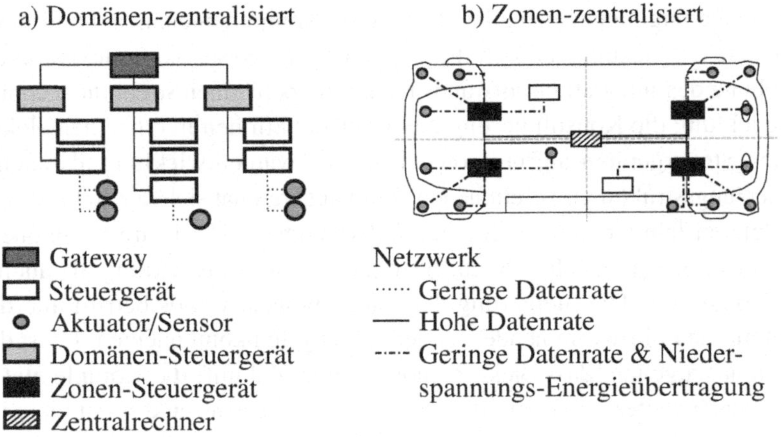

Abbildung 2: Domänen- und Zonen-zentralisierte E/E-Architekturen

Diese Dissertation hat sich zum Ziel gesetzt, eine Methodik zu erarbeiten, mit der eine solche zonale Elektrik/Elektronik-Architektur unter Optimierung des Kabelstrangs entwickelt und in bestehende Packagekonzepte integriert werden kann. Durch die geometrische Strukturierung, die zunehmende Vernetzung innerhalb der Zone und abnehmende Vernetzung zwischen den Zonen sowie die service-orientierte Architektur als gestaltungs-leitende Logik in der Zone wird eine Architektur entwickelt, welche die Komplexität für die Entwicklung und Integration reduziert und die Wiederverwendbarkeit steigert. Anhand dieser drei Aspekte wird das Gesamtsystem in mehrere kleine Systeme, die Zonen, geteilt, wodurch der Integrationsaufwand sinkt sowie der Entwicklungsaufwand durch die Wiederverwendbarkeit von Zonen bzw. von Teilen der Zone gesteigert wird. Das Package gewinnt im Zusammenhang mit zonen-zentralisierten Konzepten an Bedeutung, da diese die Fahrzeuggeometrie als Strukturierungskriterium verwenden. Es gibt bereits Verfahren, die sich mit der optimalen Allokation von Software zu Hardware oder der kostenoptimierten Auslegung der Batteriegröße sowie deren Anzahl beschäftigen. Diese Verfahren wurden zwar für bereits bestehende Architekturkonzepte entwickelt, werden aber auch weiterhin für Zonen-zentralisierte Architekturen benötigt. Ein Verfahren zur Aufteilung und Bestimmung der optimalen Zonenanzahl wird hingegen erst für Zonenzentralisierte Ansätze erforderlich und ist demzufolge noch nicht vorhanden.

Das hier vorgestellte Verfahren ist in Abbildung 3 anhand der wichtigsten Schritte dargestellt. Die Vereinfachung des Kabelstrangs ist einer der stärksten Motivatoren für den Wandel hin zu zonalen Architekturen, demzufolge muss dieser in der Systemmodellierung berücksichtigt werden. Da die Optimierung von zonalen E/E-Architekturen mittels geometrischer Kriterien erfolgt, ist eine kompatible Modellierung des Fahrzeugpackagings und des Kabelstrangs notwendig. In anderen Architekturparadigmen, wie beispielsweise der Domänenzentralisierten Architektur, bietet die Berücksichtigung der Fahrzeuggeometrie im Optimierungsprozess keinen Mehrwert für die Gestaltung der Architektur selbst. Ein Volumenmodell ermöglicht die Modellierung der geometrischen Zonen, wohingegen das Graphenmodell den Kabelstrang abbildet. Durch die Kopplung gemeinsamer Parameter und Randbedingungen wird sichergestellt, dass beide Modellierungsansätze kompatibel sind. Daher wird zunächst die bestehende Fahrzeuggeometrie auf ein Gitter aus Kanten und Knoten reduziert. Zum einen wird das Fahrzeug durch das Gitter in individuelle Volumenelemente (Voxel) unterteilt, welche eine Bauraumbelegung ermöglichen. Zusätzlich kann das Voxel-Modell dazu verwendet werden, die Zonen und ihre geometrischen Grenzen zu beschreiben, indem die einzelnen Voxel einer Zone zugeordnet werden. Zum anderen kann das Gitter in einen Graphen, bestehend aus Knoten und Kanten, transformiert werden, welcher die Identifikation optimaler Kabelstrangpfade ermöglicht.

Die zonale Architektur wird unabhängig vom gewählten Modellierungsansatz definiert und kritische Hardwarekomponenten werden modelliert. Da Zonenzentrierte Architekturen aktuell noch keine Anwendung in Serienfahrzeugen finden und hauptsächlich Gegenstand der Forschung sind, variiert das Verständnis des Systems und seiner Grenzen stark. Es wurde ein semiformales Systemmodell für zonale E/E-Architekturen entwickelt. Dieses berücksichtigt sowohl das Fahrzeugpackaging als auch den Kabelstrang und die Separierung in Zonen. Die Implementierung ist darin nicht definiert und einzelne Aspekte sind optional, um eine universelle Anwendbarkeit zu gewährleisten.

Die Optimierung des Kabelstrangs wird auf einzelne *Single-Source-Shortest-Path* Probleme reduziert. Dazu werden in einem ersten Schritt alle Pfade

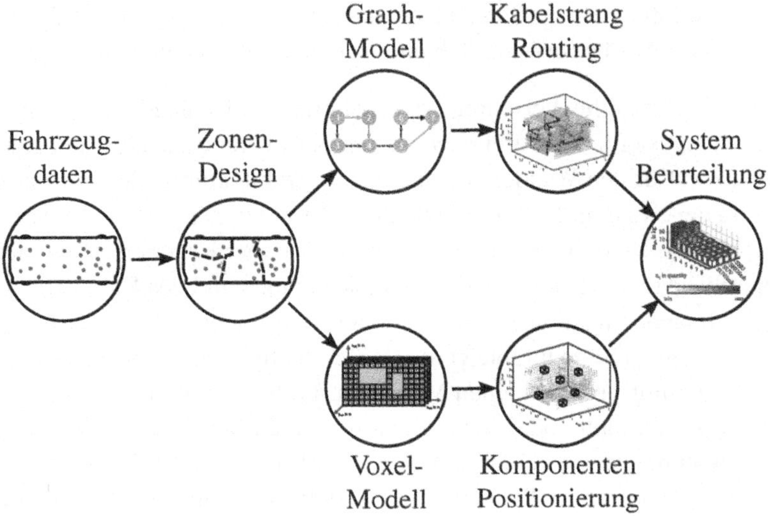

Abbildung 3: Übersicht der entwickelten Methode

hinsichtlich ihres kürzesten Wegs im Graphen optimiert. In einem zweiten Schritt wird eine Bündelung dieser Pfade bei minimaler Steigerung des Aufwands angestrebt. Der Aufwand wird anhand der Pfadkosten bzw. Kantenkosten gemessen, welche durch ein Kostenkennfeld modifiziert werden. Das Kostenkennfeld wird abhängig von Anforderungen wie beispielsweise präferierter Pfade oder Temperatur gestaltet. Zusätzlich zur Zonen- und Pfadoptimerung werden die Hardwarekomponenten der E/E-Architektur anhand von Regressionsmodellen modelliert und in das Voxel-Modell eingebunden.

Es wird eine Fallstudie mit einem Fahrzeug der Plattform, welches die Entwicklung dieser Methode motiviert hat, und einem Serienfahrzeug durchgeführt. Zur Bewertung wurden Kriterien zur Beurteilung der physikalischen Eigenschaften sowie der Komplexität der gesamten Architektur und des Kabelstrangs festgelegt. Die Untersuchungen zeigen, dass die optimale Zonenanzahl stark vom Fahrzeugmodell sowie den gewünschten Anforderungen abhängen. Sowohl die Fahrzeuggeometrie als auch das Antriebs- und Batteriekonzept führen zu signifikaten Unterschieden. Modifikationen wie beispielsweise die

Türposition resultieren in kleinen Abweichungen, deren Kenntnis jedoch ebenso wertvoll bei der Konzeptbewertung und -auswahl ist.

Die Methode zeigt anhand von zwei Fallbeispielen das Potential für die Optimierung zonaler E/E-Architekturen bezüglich des Kabelstrangs auf. Sowohl hinsichtlich der physikalischen Eigenschaften als auch der Komplexität zeigen sich maßgebliche Unterschiede. Durch die Flexibilität bei der Genauigkeit der Eingangsdaten kann die Methode bereits in einer sehr frühen Phase des V-Modells, beispielsweise der Systemdefinition, unterstützen. Bei zunehmender Kenntnis über die Fahrzeugsysteme und ihrer physikalischen Eigenschaften können Konzepte verschiedener Varianten so innerhalb weniger Stunden bewertet und verglichen werden, um die Effizienz der Systementwicklung zu verbessern. Auch in späteren Phasen des V-Modells, beispielsweise wenn bereits Prototypen entwickelt werden, kann durch die Methode der Einfluss von Parameteränderungen ermittelt und somit die Effizienz des Prozesses sowie die Qualität des Systems gesteigert werden.

Abschließend wird die Vereinbarkeit mit bestehenden Architekturkonzepten sowie die Übertragbarkeit auf andere mechatronische Systeme diskutiert. Zwar ist die Vereinbarkeit gegeben, es bleibt jedoch abzuwarten, ob die Fahrzeughersteller Vorteile in einem schrittweisen Übergang sehen oder eine schlagartige Einführung der zonalen E/E-Architektur mit der Einführung neuer Plattformen kombinieren. Die Übertragbarkeit ist prinizipell gegeben, die Systeme müssen jedoch gewisse Vorraussetzungen erfüllen. Diese sind eine starke räumliche Verteilung von elektronischen Komponenten sowie eine intensive Vernetzung zwischen diesen. Beides resultiert in einer hohen Komplexität des Kabelstrangs, welche schlussendlich den Wechsel zu einer zonalen E/E Architektur und somit auch den Einsatz dieser Methode gewinnbringend macht.

Abstract

Today complexity of cars is highly dependent on the point of view. While the market notice it as a machine that basically is able to drive and offers some comfort functions, technical development realize it as a compound of various sub-systems with high dependencies. This is due to the fact that a lot of design, coordination and verification actions are required until all systems work together consistently in the final product. These actions are necessary, because for the realization of each customer recognizable function, often called feature, many individual functions must be developed and combined. Different disciplines, such as electronics, mechanics and software are closely linked in an environment of high cost and time pressure. Naturally this leads to an area of conflict, where it is essential to identify the best economical and technical compromise satisfying the customer. A central concept here is integration, i.e. the incorporation of technological systems, considering technological interfaces and various interest groups. The identification of this compromise often takes place in an interdisciplinary multi criteria solution space.

The Electric/Electronic architecture (EEA) of today's vehicles is the system handling the entire data and energy transmission. A few decades ago, this involved a handful of components, whereas today it concerns over 100 electronic control units as well as numerous actuators and sensors. However, not only has the sheer number increased, the complexity has also risen due to new functions and increased interconnectivity. Complexity can be understood as a measure of the systemic difficulty of a system and the interactions between the system components. It therefore increases as the number of components and interactions increases. This term should not be confused with complicacy, as this is a measure of the difficulty in understanding a system or cause-effect relationship. For technical development this means a multiplication of the effort involved as the complexity of the interaction, both within the system and with other systems, increases exponentially, due to advancing digitalization and networking as well as greater automation of driving functions.

Since the number of system elements and their interactions is constantly increasing, the complexity of integrating system components is also constantly rising. Current E/E architectures were developed under different circumstances, i.e. a lower number of components and integration depth, which means that the increase in system complexity results in an additional integration effort. On the one hand, developers are striving to consolidate many functions in individual electronic control units in order to simplify both software development and the networking structure. On the other hand, the wiring harness has become one of the most complex hardware components in the vehicle in recent decades, both in terms of development and production. These trends are pushing current EEAs to their limits, as the increasing complexity is barely manageable. It can be assumed that these trends will continue to increase the complexity of vehicles in the future, consequently current architecture concepts will not be suitable anymore.

One architecture concept that is designed to solve these problems is the zonal EEA. The core of the idea is not to structure the architecture according to technological criteria, e.g. transmission technology resp. protocol, or according to functional criteria, e.g. powertrain, comfort, infotainment. Instead, the vehicle is divided into zones that are defined on the basis of spatial characteristics. Domain centralization made it possible to reduce the need for communication, as a large share of this took place within a domain and was thus limited to this domain. As a result of increasing networking, cross-domain communication continues to grow, pushing the existing concept to its limits. With zone centralization, the majority of communication should take place within a zone and inter-zonal communication is to be implemented using a network operating at a high data rate. Coupled with a service-oriented software architecture, the associated software can be developed more independently of the hardware. A decisive advantage is the reduction in complexity of the wiring harness, which is facilitated by splitting it into multiple zone wiring harnesses. In addition, the modularity of hardware and software increases, as components can be reused in other zones.

The aim of this dissertation is to develop a methodology that enables such a zonal EEA to be developed and integrated into existing vehicle package concepts by optimizing the wiring harness. Through geometric structuring, increasing networking within the zone and decreasing networking between the zones as well as the service-oriented architecture as the design-guiding logic in the zone, an architecture is developed that reduces the complexity for development and integration and increases reusability. Based on these three aspects, the overall system is divided into several small systems, the zones, which reduces the integration and development effort through the reusability of zones or parts of the zone. The package is becoming increasingly important in combination with zone-centralized concepts, as these use the vehicle geometry as a structuring criterion. There are already methods that deal with the optimal allocation of software to hardware or the cost-optimized design of the battery size and its number. These methods have been developed for existing architecture concepts and are still required for zone-centralized architectures. However, a method for dividing up and determining the optimum number of zones is only required for zone-centralized approaches and therefore is not available.

The simplification of the wiring harness is one of the strongest motivators for the change to zonal architectures, so this must be taken into account in the system modeling. Since the optimization of zonal E/E architectures is based on geometric criteria, a compatible modeling of the vehicle packaging and the wiring harness is necessary. In other architecture paradigms, such as domain-centralized architecture, the consideration of vehicle geometry in the optimization process offers no added value for the design of the architecture itself. A solid model enables the modeling of the geometric zones, whereas the graph model represents the wiring harness. Coupling common parameters and boundary conditions ensures that both modeling approaches are compatible. Therefore, the existing vehicle geometry is first reduced to a grid of edges and nodes. On the one hand, the grid divides the vehicle into individual volume elements (voxels), which make it possible to allocate installation space. In addition, the voxel model can be used to describe the zones and their geometric boundaries by assigning the individual voxels to a zone. On the other hand, the grid can be transformed into a graph, also consisting of nodes and edges, thus enabling the identification of optimal cable paths.

Regardless of the modeling approach chosen, the zonal architecture is defined and critical hardware components are modeled. Since zone-centralized architectures are not yet used in production vehicles and are mainly the subject of research, the understanding of the system and its boundaries varies greatly. For this purpose, a semi-formal system model for zonal E/E architectures is designed using SysML. It takes into account the vehicle packaging as well as the wiring harness and the zone separation. The implementation is not defined in it and individual aspects are optional ensuring universal applicability.

The optimization of the wiring harness is reduced to individual Single-source shortest path (SSSP) problems. In a first step, all paths are optimized with respect to their shortest path in the graph. In a second step, the aim is to bundle these paths while minimizing the effort. The effort is measured using the path costs or edge costs, which are modified by a cost map designed depending on requirements such as preferred paths or temperature. In addition to zone and path optimization, the hardware components of the EEA are modeled using regression models and are integrated into the voxel model.

Subsequently, a case study is carried out with a vehicle of the platform that motivated the development of this method and a production vehicle. Criteria for evaluating the physical properties and complexity of the overall architecture and wiring harness are defined. The investigations show that the optimum number of zones depends heavily on the vehicle model and the desired requirements. The vehicle geometry as well as the drive and battery concept lead to significant deviations. Modifications such as the door position result in minor differences, but knowledge of these is also valuable for concept evaluation and selection.

The method uses two case studies to demonstrate the potential for optimizing zonal EEAs with regard to the wiring harness. There are significant differences in terms of physical properties and complexity. Due to the flexibility in the accuracy of the input data, the method can provide support at a relatively early stage of the V-model, for example in system definition. With increasing knowledge of the vehicle systems and their physical properties, concepts of different variants can be evaluated and compared within a few hours to ensure the efficiency of the system development. Even in later phases of the V-model, for example when prototypes are already available and being tested, the method can be used to determine the influence of parameter changes and thus increase the efficiency of the process and quality of the system.

Finally, compatibility with existing architecture concepts and transferability to other mechatronic systems are discussed. Compatibility is given, but it remains unclear whether vehicle manufacturers see advantages in a gradual transition or combine an abrupt introduction of the zonal EEA with the introduction of new platforms. Transferability is possible in principle, but the systems must meet certain requirements. These are a high spatial distribution of electronic components and strong interconnection between them. Both result in a high complexity of the wiring harness, which ultimately makes the transition to a zonal EEA and therefore the use of this method beneficial.

1 Introduction

Products and their systems do not only need to fulfill one criterion, in most cases they are designed in a trade-off between multiple aspects. As an example, one can think of a vehicle traction battery, that should have a low weight, high energy content, temperature and vibration resistance and a long lifetime. It is not possible to fulfill all these completely, therefore technical development needs to identify the best compromise in this area of conflict. There are two routes to resolve such a trade-off, the first is the optimization with respect to multiple criteria and the second is the development of innovations. The same applies to the transportation sector, which is affected by various challenges such as automated driving, digitization and global warming. While global warming motivates the transportation industry to build more sustainable vehicles using an electric powertrain, automated driving and digitalization require new functions and tighter coupling between the vehicle systems [3]. These trends pose major challenges for the automotive industry, which has resulted in a fundamental transformation of the vehicle and its systems [4]. Consequently, the rising number of system components and their increasing interdependence has shifted vehicle design from a hardware-centered to a software-dominated approach, today often referred to as a software-defined vehicle. As a result, the complexity of vehicles continuously rises, making integration of components more and more challenging [5]. In Figure 1.1, a visualization of the projected transformation by the market size of characteristic hard- and software elements in the years 2020 to 2030 is shown. The market share of sensors is expected to grow, but other components will also experience a significant gains, albeit unspecific electronic components and Electronic control units (ECUs)/Domain control units (DCUs) will have lower growth. This indicates the previously mentioned transformation towards software-defined vehicles through a rising share of electronics, which is necessary for the realization of digital functions. This circumstance results in an increase in technical complexity and the associated engineering design tasks.

J. Maier, *Optimization of Zonal Electric/Electronic Architectures Using a Coupled Modeling Approach of Package and Wiring Harness*,
Wissenschaftliche Reihe Fahrzeugtechnik Universität Stuttgart,
https://doi.org/10.1007/978-3-658-48188-9_1

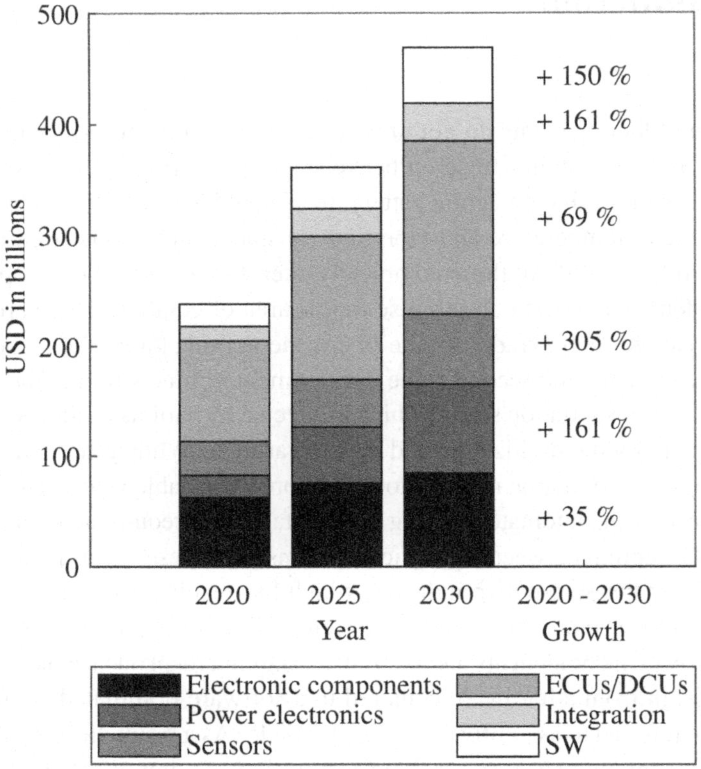

Figure 1.1: Trends in automotive electronics and software adapted from [6]

The Electric/Electronic architecture (EEA) of a vehicle is responsible for the in-vehicle data and energy transmission and thus is a vital part for handling the complexity of vehicles and their architectures [7]. It is clear, that the EEA as a highly interconnected vehicle system awaits significant challenges [8]. The increasing amount of hardware and software components will continuously increase the vehicle complexity and by this push existing EEA to its limits and beyond [9]. Consequently the improvement of an existing systems is not adequate anymore to tackle this transformation, instead innovation by means of zonal EEA is inevitable. A barrier for the realization of zonal EEAs is the lack of methods for their design, especially with regard to the optimization of the hardware architecture, considering new geometric design requirements.

In the last two decades most vehicle manufacturers have shifted from decentralized to domain-centralized EEA. The bundling in domains follows functional criteria, reducing complexity as the intra-domain interdependence has increased over the years. Today manufacturers are again conflicted between extending the existing or redesigning a new architecture, which is suitable for future vehicle technologies. Domain-centralized architectures brought many improvements, but there are some disadvantages that will evolve into bottlenecks by the transformation of the automotive industry. Zonal EEAs are seen as the solution for the problems arising from the perpetual extension of the existing architecture. According to Burkacky et al., the simplification of the ECU landscape will bring the most benefit; moreover they mention four major factors that should be evaluated when technical solutions are defined, namely [10]:

- Facilitation of secure over-the-air updates
- Decoupling of hardware and software
- Increasing silicon consolidation and integration
- Reduction of wiring harness complexity

The first and third factor are intrinsically realized by the architecture shift, therefore these two are not the focus of this thesis. The second, decoupling of hardware and software, is not implemented by the architecture shift itself, but it offers great potential for utilization. The decoupling of both enables faster development and adaption as well as reconfiguration of functions in static cases, for example, for updates or dynamic cases such as reallocation in safety-critical states. The last of the four factors is basically realized by the architecture shift, and holds significant potential for optimization. Zonal EEA, respectively bundling of ECUs in Zone control units (ZCUs), enables splitting of the wiring harness and assignment in zones of individual sub-modules leading to more manageable subsystems [11]. In addition length, mass and complexity can be reduced, but it requires the transmission of higher data rates, which is tackled by the progress of automotive Ethernet [12]. Furthermore it must be clear, that zone-centralization does only reduce the complexity of the system and the wiring harness if both have a certain level of complexity. For system, respectively vehicles, having a low number of functions and components, it is doubtful if zone-centralization will ever bring a benefit for them. Consequently it is the aim of this thesis to develop and evaluate a method able to identify an optimized electric and geometric design for a zonal EEA, leading to improvements for the architecture itself and the wiring harness. The scope is limited to the hardware architecture of zonal EEA, comprising the electrical design and the geometrical integration with respect to software and technology boundaries that are already known today.

1.1 Problem and Research Questions

Crucial parameters for the design of these architectures are the number of zones as well as the position of the data and energy hubs, in the following called ZCUs, and the assignment of electronic components to those. These parameters are decisive for the design of the wiring harness as they influence the physical characteristics as length and mass as well as the structural design. By optimizing the ZCU position and the component assignment it is possible to reduce the wiring harness complexity and utilize the potentials of this architecture. This results in a reduction of development and production effort of both, EEA and wiring harness. Focusing on the number of zones and ZCUs, there are conflictive targets that need to be fulfilled. On the one hand, higher zone numbers will reduce the complexity through modularity and simplification, on the other hand will rising zone numbers increase the number of components and their cost. For specific zone numbers, it will even result in higher complexity due to the additional effort of orchestrating the individual zones. The combination of these aspects becomes relevant by the introduction of zonal EEA, hence there are no existing methods for this specific aspect. Relevant methods optimize aspects such as packaging, traction batteries or wiring harnesses isolated, as this was meaningful for domain-centralized methods. Nevertheless they are not instrumental in the identification of zonal EEA design. The development of the method is inspired by the design process of the UNICARagil architecture. As this architecture and use-cases include future concepts and technologies, they reflect the prospective requirements EEAs have to cope with. In the course of the UNICARagil project, a common vehicle platform for four use cases was developed to tackle the increasing complexity and future trends. While one of the aims was to develop a zonal EEA, it was not clear how to design the architecture with respect to the new concept. The number of zones, their design and the configuration of the wiring harness were the main challenges. As there was no method that supported the identification of solutions to this aspects, intuitive and practical approaches were selected. The vehicle is divided into four symmetrical zones separated by the longitudinal and transversal axis of the vehicle, justified by the four power intensive wheel-hub drives, but the wiring harness was neither evaluated nor optimized. While the solution basically worked, the process as well as the final implementation revealed shortcomings.

Based on the practical experience in the design of the zonal EEA for the UNICARagil architecture, the following research questions were derived:

- Question 1: "How many zones are suitable for a given use case?"
- Question 2: "How should these zones be designed to enable a package compatible design?"
- Question 3: "How can the complexity of a wiring harness be determined to evaluate its quality in the solutions found?"

Question 1 targets the aspects that the stakeholders of this technology have completely different understanding of ZCUs and zones leading to a largely varying number of zones. Furthermore, this number is in most cases not justified neither by theoretical knowledge of the system nor by practical reasons. Consequently, there is a lack of a common understanding of the system and corresponding evaluation criteria, which both need to be established.

Question 2 connects the theoretical analysis with the practical implementation of the system. After establishing a common understanding of zonal architectures, finding solutions for its geometric aggregation with existing systems and integration into vehicles is essential. The already high complexity of the spatial design in vehicle is further pushed by the crucial role of the data and energy hubs. Therefore, a method to realize the transition from system definition to package concepts is necessary.

The complexity of the wiring harness is a crucial aspect for the identification of the suitability of EEAs. However researchers struggle defining complexity of technical systems in general, but particularly in the context of wiring harnesses. Thus question 3 targets the definition of complexity, which is a rather elusive term.

The three research questions imply the following hypothesis: *There is an interdependence between the geometrical design of the zones as well as their number and the properties of the EEA and the wiring harness.*

The aim of collectively answering these three questions is the development of a method for the design of a zonal EEA that is compatible with an existing vehicle package. Based on this, the influence of the zone design on the properties of the EEA and the wiring harness is analyzed and the hypothesis discussed.

1.2 Methods and Findings

There are no methods that address exactly the issue described in the previous chapter. When isolating individual aspects of the problem, there are various methods that find solutions to these. The problem is reduced to three individual aspects that need separate solutions and a common model to combine them. The three core aspects are the traction battery configuration, the wiring harness routing and the zonal design. Traction batteries are optimized with respect to diverse criteria using genetic algorithms [13, 14]. These approaches present solutions for the multi-criteria problems of battery design, but do not consider the architecture level or the geometrical design of the vehicle. The genetic algorithm in this thesis is adapted to consider only aspects relevant for the zonal design enabling a more efficient process. The wiring harness routing is simplified as a SSSP, which has many practical and theoretical solutions as optimal path planing is an issue occurring in many disciplines. The benefit of the procedure introduced here is the package specific consideration of requirements and the cable bundling facilitating a more detailed modeling. As there are no methods for the zonal design, a k-means clustering based approach is developed to identify the optimal data and energy hubs as well as a component allocation algorithm.

Consequently, a holistic method for the development of zonal EEAs is developed by combining a k-means clustering based approach for the zone design with adapted existing approaches for battery, package and wiring harness design. This contributes to close the existing gaps in two ways. On the one hand it offers an optimized zonal EEA design and due to the interdisciplinary characteristic of the method it additionally simplifies the aggregation of the electrical system design with the geometrical integration of the EEA itself and of the entire vehicle. On the other hand it permits a more efficient development process as it specifies its elements and their sequence. In addition it engenders the target-oriented development on zonal EEAs by identification of purposeful variants even under uncertainties.

1.3 Research Methodology and Thesis Structure

The research of this thesis follows the principle of action research, although initially defined for social science it offers a broadly applicable method for various disciplines [15]. Action research combines theory and practice to identify practical solutions for real world problems. For this purpose an existing problem is diagnosed and corresponding actions are planned, taken and evaluated [16]. In the evaluation these actions, respectively the practical solutions, for the research object are reviewed, which leads to a greater understanding regarding the theoretical knowledge about the problem. This process can be continued, as what is learned will either change the research object or the theoretical knowledge of the problem or both. Answering each of the three questions led to a change of the problem itself, resulting in a new understanding and enabling new actions. This cycle spanned by action research is visualized in Figure 1.2, wherein the numbers in the small circles indicate the chapter focusing on the corresponding activity. In Chapter 2 the EEA system is introduced, interdependence between subsystems, as well as with its environment are analyzed and diagnosed. Chapter 3 embodies the action planning by containing a discussion of already existing methods and evaluating their suitability as well as applicability. Based on the action planning, an action was taken by developing a method aiming to tackle the issue in zonal EEA design, which is introduced in Chapter 4. The consequences of the action are evaluated in Chapter 5 by means of two vehicles, one is the autoSHUTTLE, which was the trigger of this action research, the other is the model of a series vehicle. To achieve greater understanding and deepen theoretical knowledge, the results of the evaluations are discussed in Chapter 6. This comprises a critical interpretation of the results under consideration of approximations, context and research object to deepen the knowledge about effects of the actions and the aspects future work can address. The context of this research is the collaborative project UNICARagil of technical universities and companies of Germany. Due to project targets and limitations as well as stakeholder interests, some design decisions represent the best compromise for the entire vehicle and not the EEA, or simple were fixed from the start due to administrative reasons.

Figure 1.2: Canonical action research cycle and their assignment to the thesis chapters adapted from [17]

2 Vehicle Electric/Electronic Architectures

The aim of this chapter is the definition of basic terms and concepts as well as the state-of-the-art. In the following a description of different system concepts is given, which are substantiated by the exemplification of its layers. Based on this, the system is contextualized in the entire development process and subsequently the research gap is identified.

Vehicle EEAs have evolved over the last decades leading to varying definitions. In the 1970s to 1990s, the term was used rather rarely, instead the system was named powernet. At that time its main function was to supply electric power to all Low-voltage (LV) electronic components, as data exchange was mostly conducted between two ECUs via a single cable. However, the growing number of ECUs and their interconnectivity required the introduction of data buses, enabling the transmission of a larger amount of data between several ECUs. Today, both data and energy transmission are inseparable as most components require both and the cables follow similar paths. The wiring harness is the aggregation of these and further cables, composing one of the most complex components in today's vehicle development. Since the definition of the EEA varies among engineers and researchers, it is important to clarify the term for this dissertation.

Definition 1. *The* ***electric/electronic architecture*** *of a vehicle is the aggregation of all subsystems enabling storage or transmission of data and power in the vehicle. It is comprised of hardware and software components as well as devices for controlling and monitoring these tasks.*

Energy storage is traditionally assigned to the vehicle power supply. The shift to electromobility has changed its role, as the High-voltage battery (HVB) is essential to the electric powertrain; thus it is considered as part of the powertrain system when designing those. Nevertheless, the HVB is necessary as an energy storage for all electronic components, consequently the terms EEA and power supply system in this dissertation comprise the HVB. It is necessary to carry out most of the cars' functions and is spread out all over it, both leading to a large number of interfaces on the functional and packaging layer [18].

J. Maier, *Optimization of Zonal Electric/Electronic Architectures Using a Coupled Modeling Approach of Package and Wiring Harness*,
Wissenschaftliche Reihe Fahrzeugtechnik Universität Stuttgart,
https://doi.org/10.1007/978-3-658-48188-9_2

The aim of this chapter is to deepen the understanding for these interrelations and highlight the impact of the technology shift in the automotive industry. Therefore, previous and future architecture concepts are introduced and a detailed explanation of the layers is given. Based on this, the role of EEA design in the vehicle development process is discussed to highlight the contribution of this thesis.

2.1 Previous and Future Concepts

The EEA is currently undergoing a fundamental change, on the one hand it is driven by the high and continuously increasing number of ECUs and functions, on the other hand by the established *One Function per ECU* approach. The architecture approaches discussed in this section are also available in the other industries, as computer networks are necessary for many application. The solution found in these other industries are not directly transferable to the automotive industry due to different boundary conditions such as low cost and weight. For systems of low complexity, a decentralized architecture offers advantages in terms of communication traffic, variant diversity and upgradeability. Higher complexity turns this into disadvantages, since the development effort and the communication between the ECUs is enormous due to a high number of interfaces. In this work, complexity is understood as the effort that is required to handle a system. Kreimeyer defines the following six aspects that characterize complexity [19]:

- Structure: Structuring criteria that define and limit groups
- Configuration: Arrangement of the individual components
- Interaction: Actions and states of the individual components
- Interference: Transferability of the behavior and structure of the components to those of the system
- Response: Components may adjust due to changes of adjacent parts
- Understandability: Effort to understand and verify the system

Whenever there are more structuring criteria, configurations, interactions or a greater response, the complexity of a system rises. It also increases, when there is low interference and understandibility. Albeit is should be strived for low complexity when designing a system, constraints and requirements and targets drive the system towards high complexity. In software-driven systems, such as automobiles, the number of variants is high. The cause is the shorter development time of software compared to hardware resulting in more variants over the entire product cycle. Using centralized architectures can make the increasing number of variants manageable, as a result of simplified hardware development and a reduction of interfaces, which additionally reduces variant numbers. In centralized architectures communication is coordinated by several central ECUs, which are crucial to ensure real-time requirements for high data loads [20]. Since Original equipment manufacturers (OEMs) are bound to their product development cycles, a gradual transition from decentralized to centralized architecture is inevitable. Figure 2.1 provides an overview of the most relevant EEA architectures. Decentralized architectures have been used in many vehicles that were produced in the past decades, while most EEAs today are domain-centralized or use a combination of decentralized and domain-centralized elements. Zonal centralization is being discussed as next crucial step in EEA centralization, consequently it has been and is intensively researched, but an application in series-vehicles has not yet taken place [21–23].

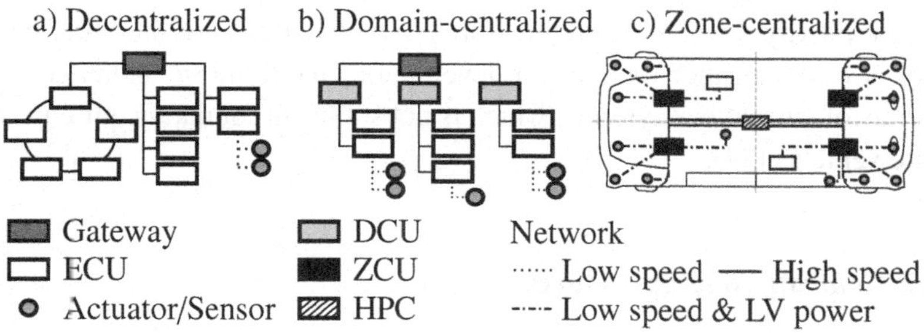

Figure 2.1: Decentralized, domain-centralized and zone-centralized EEA

2.1.1 Decentralized Architectures

In decentralized architectures many functions are implemented separately in one ECU, and the connections between them and other components are made via individual communication paths. This procedure has been used for decades resulting in a high number of electronic nodes, in high-class vehicles there are up to 100. Consequently, the number of interfaces and cables has also increased, leading to rising cost. Data buses reduce the number of cables by enabling the distribution of information to all ECUs connected to the bus. Therefore, distance and number of linking elements can be minimized as only one of them is necessary to connect an electronic component to the data bus, instead one for each connection between two ECUs. By using different bus systems, it is possible to adapt the data rate and safety mechanisms to the information type. In the past, these approaches helped to reduce the cost of cables as well as the communication effort and to manage complexity. Clearly defined interfaces make it straightforward to combine different versions for implementation and testing during development. Diagnosis and replacement of individual components is facilitated by a limited number of functions in one piece of hardware. Due to the continuously growing share of software in vehicles, the number of ECUs, cables and data traffic is increasing. Consequently, interface quantity and (software) variants have experienced a rise, however concurrently time to market and thus development time decreases. On the one hand this drives costs and complexity for vehicle architectures to new heights, on the other hand it poses new technical challenges for developers. One of those challenges is the data rate of current bus systems, which is not sufficient for data requirements of future vehicles [24].

2.1.2 Centralized Architectures

Centralization implies the consolidation of multiple ECUs into one piece of hardware, unifying their functions and software components. While aggregation reduces the number of ECUs, the resulting hardware requires higher computing power and more powerful communication modules [25]. Centralization does not only take place at the software-hardware allocation level, it is a comprehensive attempt that covers most hardware and software components as well as the vehicles geometric design. While domain centralization is widely being adopted as state-of-the-art and is gradually being implemented, zone centralization still requires many issues to be solved. Various companies have already shown concepts, but these have not yet been implemented in production vehicles. In addition to these two centralization approaches, other concepts such as a central architecture with a powerful ECU or outsourcing functions to the cloud are being discussed [26].

Domain-centralization structures the functions of the vehicle into domains, each comprising highly interrelated functional groups of the vehicles. While each original equipment manufacturer or supplier specifies the domains individually most definitions embrace the domains AD and ADAS, infotainment, passenger compartment, powertrain and safety [27]. Domains require a DCU, its associated ECUs as well as further electronic components such as actuators and sensors [28]. The DCUs come with high computing power, because they bundle and process the functions of their corresponding domain and interact with other DCUs [29]. ECUs are designed for specific functional scopes and, together with the sensors and actuators assigned to them, form smart sensors and actuators. It is assumed that the connectivity of functions in vehicles will steadily increase, leading to high dependency and communication overhead between domains. Consequently, the boundaries between domains become blurred and there can be a shortage of bandwidth. An approach to tackle these issues is the cross-domain-centralized architectures, combining several domains in one ECU to alleviate the problem of communication overhead. Nevertheless, this architecture concept may reach its limit, considering the aspiration for automated driving and digitalization. Approaches to tackle these boundaries are virtual ECUs, cross-domain centralization and zone centralization. Virtual ECUs enable to shift software functions from a specific ECU to another and cross-domain centralization comes with consolidation of highly interconnected domains into one.

In addition to increasing complexity, the growing number of functions and ECUs also lead to an increase in the accumulated cable length, as domains can extend to the entire vehicle geometry. To counteract both, the zone-centralized architecture structures the vehicle according to geometric aspects [30]. The vehicle is divided into geometric zones, each having a ZCU responsible for coordinating the actuators, ECUs and sensors of the corresponding zone. In contrast to the domain-centralized architecture, it is meaningful to integrate the electric power supply in the zonal-centralized architecture concept to benefit from shared cable paths and a common hardware. ZCUs are designed to manage not only the data traffic, but also the LV electric power distribution within their zones [31]. The wiring harness design profits significantly from this difference, as the benefits of zonal architectures are extended not only to data transmission, but also to power supply. In addition to the ZCUs, there is a central vehicle ECU that coordinates them. The exact distribution of computing power between ZCUs and central ECU is not yet clear. However, this will largely determine the amount of processed data in a ZCU in comparison to the data transferred to the central ECU. A cloud backend can extend this architecture, allowing functions to be outsourced or access to a database. OEMs and suppliers expect zone-centralized architectures to reduce the weight of the wiring harness by 15 % - 20 %. Implementations in production vehicles are currently unknown, but there are concepts and initial approaches such as in the prototype platform of the UNICARagil vehicles [32–35].

2.1.3 Requirement Shift

So far this chapter focus has been on the architecture concepts of the past, present and future. The following section strives to highlight the cause of this shift in the automotive industry, which has been undergoing a transformation for several years [36, 37]. The drivers of this change are the increasing demand for automated, flexible and sustainable mobility in an environment that is subject to increasing digitalization [38–40]. Considering these aspects as being isolated is not meaningful, as the transformations complexity is caused by their interdependence. Automation of driving functions requires further ECUs and new sensors for data processing and environment detection [41]. Therefore, the power demand increases and at the same time the complexity of the system rises. Digitalization covers the development as well as the vehicles features. While customers wish for greater digitalization and connectivity, developers need a more digital development process in order to be fast and flexible enough to engineer competitive vehicles [42]. This requires manufacturers not only to develop new features, but also to reinvent development processes and tools.

The desire for sustainability supports the spread of Electric vehicles (EVs) and elimination of vehicles with internal combustion, significantly changing the structure of the EEA [37]. New components and concepts are required to realize EVs, such as higher voltage levels and traction batteries. The high power loads, such as electric drives, must be operated at High-voltage (HV) levels to reduce losses, while many existing loads still require LV. The amount of transmitted electric power increases drastically, resulting in new challenges for the wiring harness design. As an outcome the space demand for the electronic components has risen as well as due to stricter safety precautions when using HV components. Overall, electrification of the Powertrain (PT) demands a higher weight and volume share on the vehicle. In combination with automation, a high reliability for power supplies at all voltage levels is inevitable. To conclude this section, today's trends are driving a fundamental requirement shift in the automotive industry, comprising hardware and software. It is not only necessary to improve existing systems, but to apply new concepts such as zone-centralized EEAs.

2.2 Architecture Layers

The EEA is the system responsible for the in-vehicle power supply and data transmission, ensuring efficient and reliable data and energy transfer between all electronic components. While they serve the same components, one can draw a distinction between data and energy transmission as they differ in their transmitting medium and source [43]. Therefore, sometimes it is meaningful to separate the system into two subsystems: Power supply system and communication system. In this dissertation, these two terms will be used when it is necessary to highlight that only one of them is relevant in a specific context. The term EEA is preferred when the featured aspects concern both subsystems or the architecture is discussed in a more general way. A layer based description of the EEA has been established. Each layer represents a specific viewpoint of hardware, product or software developer, and for some of them it is reasonable to outline the role of the two subsystems [44, 45]. The hierarchical context of these layers follow a top-down principle in accordance to the V-model illustrated in Figure 2.2. It also follows the approach of Bender, who defines the four core product development steps planning, conceptual design, functional design and technical realization [46]. While most layers are applicable to mechatronic systems, the interdependence between them in vehicles are relatively high [47]. The responsibility for data and energy transmission results in a great number of interfaces in relation to other vehicle systems. Consequently, the EEA has high dependencies between its constituent components and to other systems, making a structuring into layers inevitable. While each perspective has an individual name, the hardware layer comprises four sub layers: Network, electric circuit, wiring harness and geometry. All of them define a different aspect of the hardware; however, there are strong interrelations among them. In consideration of the architectural shift, it becomes clear, that the role of the geometry and the wiring harness layer will change greatly. They define the spatial zone layout and therefore the interdependence to the other hardware layers and the functional layer increases. This change results in a methodical gap for combining the geometric and wiring harness layer with each other as well as with the other layers. Before addressing this methodical gap, the following subsection grant a deeper insight in the layers and their interdependence.

2.2.1 Requirement Layer

The requirement layer comprises all functional and non-functional requirements [48]. These are derived from the product specifications, which are based on market research and targeted customer groups. As the product is intended to fulfill the customer's needs, the specifications must represent the driver-perceivable functions which are mostly referred to as features. The requirements are essential for all layers below as they indicate boundaries and targets of the product. For an efficient development process it is crucial to assign requirements to specific systems and to derive more specific requirements for their components. This supports not only the development, but also the verification of requirements by bottom-up testing. Consequently, the requirement layer does not cover any technical solutions, but is fundamental for their realization.

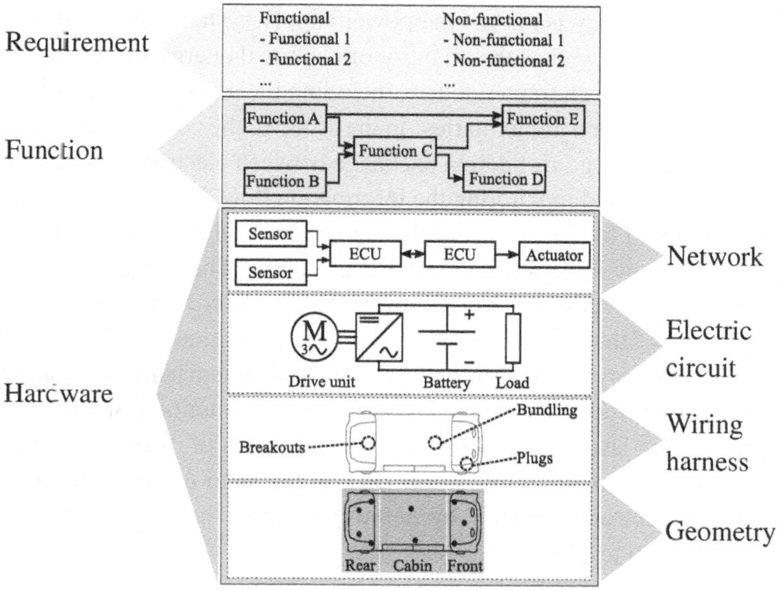

Figure 2.2: Layers in EEA development

2.2.2 Function Layer

The term function layer, also referred to as logical or software architecture, is generally understood as the structure defining the interrelation of functions in a system. The interfaces are of great importance in this context, thus interface types (input, output), data type as well as expected traffic are defined in this layer [45]. In the following section, the main functions and essential aspects for both systems are discussed.

The power supply system is designed to efficiently and reliably supply the electronic components in the vehicle. For this purpose the energy must be transported from source to sink as well as stored and released in an energy storage. The propagation of electromobility raises the relevance of the functionalities of the powernet. Previously, the LV battery served exclusively as an energy buffer during operation and provided the electrical energy for the starting process. In EVs former direct mechanically driven loads need an electrical supply and the operation strategy is part of the power supply system. Consequently, the number of loads as well as the quantity of electrical energy transferred by the power supply has risen tremendously, including the electric drives as the most prominent load as well as auxiliaries. In the scope of the communication system are the paths ensuring information exchange between electronic components as well as managing and monitoring the information flow. Due to the increased interconnectivity of functions, caused by automation and digitization, the system is essential for any in-vehicle information exchange. The expected rising share of software and electronics will put even more pressure on optimized design and operation of the system. Transmission of information is indispensable for the execution of any function that requires more than one hardware component, which applies to most functions. The relevance of functional safety has risen sharply over the last decade, especially in the context of distributed functions. The aim of the corresponding norm ISO 26262 is to ensure safe operation of complex electronic systems by defining methods and processes for their development and test [49].

2.2.3 Network Layer

The network layer determines the topology of the communication network, the signal mapping as well as the used medium, respectively protocol. The nodes may be ECUs, switches and intelligent actuators or sensors, which are interconnected via data lines [50]. In today's vehicles most data lines are bus based systems such as Controller Area Network (CAN) and Local Interconnect Network (LIN), which connect components, thereby enabling them to receive and send massages among each other [43]. Each bus system uses its specific protocol, consequently a gateway is necessary linking the diverse buses. Ethernet is rarely being used yet, but a sharp rise is expected within a few years [51]. Transmission via Ethernet enables significantly higher data rates than can be achieved for CAN and LIN, which is one of its advantages [52, 53]. Furthermore, Ethernet uses point-to point connections, which is especially meaningful for zonal architectures, but it requires switches for managing the data traffic [54].

2.2.4 Electric Circuit Layer

The electric circuit defines the electrical connections between the vehicle components, which are essential for powering components and used to transmit certain signals. The electric components can be classified into four types with respect to the electric circuit layer: Sources, sinks, transmission elements and converters. The sources provide the energy and the sinks demand energy to fulfill their intended functions. In most EVs the power source is a battery or a supercapacitor while an actuator, ECU or sensor is the sink. However, if the vehicle is in a recuperation phase, the electric motor can also be a source and the battery is a sink. Transmission elements are mostly cables made of copper with an insulating layer. For this thesis, the components of the power supply systems are classified with respect to their voltage levels. As the converters are connected to more than one voltage level, they are classified to the lowest of these voltages, except the ZCU, which is assigned to the highest corresponding voltage level. Summarizing this section, the electric circuit defines the role of all relevant components in an electrical sense as well as their connections and voltage levels.

2.2.5 Wiring Harness Layer

The wiring harness layer is used to define the physical connection between electronic components of the electric circuit and the network architecture. Therefore, it comprises the cables used to transmit data and power, including connecting elements such as pins and plugs [55]. The design and bundling of the cables is done with consideration of functional, technical as well as production specific requirements. The level of detail spans from single pins to the entire wiring harness while most crucial design parameters are the cable routing as well as their bundling. The rise in the number of vehicle functions leads to more components and greater data exchange between them, resulting in a gain of wiring harness complexity. The separation into zonal wiring harnesses can help reducing the complexity, but this requires a combined design process of the geometry and wiring harness layer that in addition ensures compatibility with the other layers.

2.2.6 Geometry Layer

The terms geometry layer, packaging and vehicle geometry are often used interchangeably and without precision. Package has a clear definition in the context of automotive engineering, whereas geometry layer and vehicle geometry are rather nebulous terms. For general purposes, the term geometry layer is more suitable, but in the context of the following chapters the term package is clearer. Both of these comprise the geometric properties of the components and their arrangement in the vehicle [56]. Aerodynamics, cabin and powertrain fundamentally define the package concept of vehicles. As the communication system has only few hardware components, there is only little interrelation to the geometric architecture. The power supply system, however, embraces more hardware, wherein especially the traction battery poses challenges for the packaging. Furthermore, it is composed of a high number of wires with varying diameters and the power converters. Although the geometry layer has a great complexity due to the high number of components and boundaries in vehicles, the architecture shift poses further challenges. The zonal design needs to be defined in this layer leading to further design goals and boundaries with the other layers.

2.3 Architecture Development Process

The EEA development process is a component focused approach. Recent trends in the automotive industry and increasing interdependence of different disciplines pose a challenge for this procedure [57]. This chapter introduces systems engineering in the context of vehicles development, highlights the gaps in existing development methods and finally points out the tradeoff in EEA design [27].

2.3.1 Development of Vehicle Systems

Due to the increasing share of electronic and software components in vehicles as well as the increasing digitalization, the term Cyber-physical system (CPS) is nowadays more appropriate when developing mechatronic systems [27]. CPS refers to systems that have a communication interface in addition to the components of a mechatronic system, which are actuators, ECUs and sensors [58]. This interface enables communication with other CPSs and the use of digital data and services. VDI 2206 specifies a system as follows:

Definition 2. *"A **system** is an unambiguously defined unit which consists of various subsystems and system elements (hardware, software, services, people, etc.) as well as their relations."[58]*

OEMs and suppliers develop mechatronic systems, respectively CPSs according to the V-model. The basic idea is to consider the customer related functions through validation and verification in all phases of product development [58]. For this purpose, development is divided into three main phases: Decomposition, implementation and integration. Decomposition and integration are drawn as legs of a V, as shown in Figure 2.3. Decomposition incorporates the specification of requirements, the partitioning of systems into subsystems or components with their corresponding requirements. Implementation comprises the realization of specifications, i.e. concept, design, and development of partial solutions. Integration embraces the combination of components and subsystems as well as the validation and verification through testing. Iterative testing and comparison with the specification (verification) and customer requirements (validation) ensures that even complex systems reliably meet customer requirements.

2.3.2 Role in the Vehicle Development Process

The entire vehicle development process follows the typical waterfall model, comprising sequential development activities that build upon artifacts achieved in previous steps. The V-Model is an iterative procedure for the development of CPS and is applied to the development activities concerning the system definition, implementation and test. The V-Model can be aligned to the overall product development process using the artifacts of the individual development

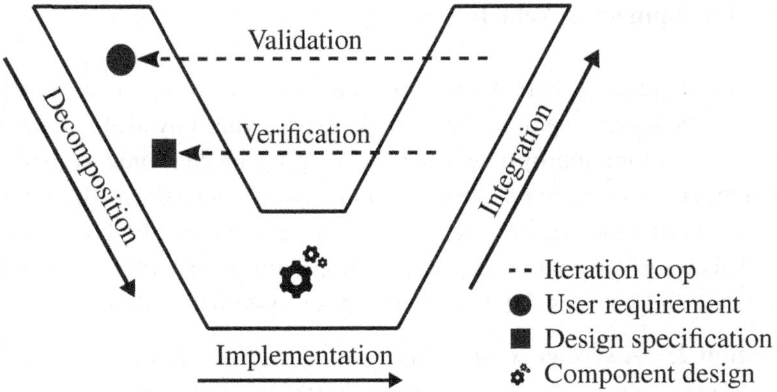

Figure 2.3: V-Model development process adapted from [58]

steps. Agil methods are not discussed in detail, as they are especially useful when developing software.

Systems engineering and the V-model are standard approaches used in the vehicle development process, but new technologies require adaption of existing or the development of new methods. For decentralized and domain-centralized EEAs, the loosely linked development of the EEA and the packaging is possible, although this results in a challenging integration task [27]. The integration of zone-centralized EEAs requires cooperative development of packaging and EEA to utilize its full potential. While many methods and tools can be adapted, the determination of the zone number and their design are completely new questions that require the development of additional methods and tools. The aim of the method proposed in this work is to support developers designing zonal EEAs. Therefore, the integration in the V-Model is necessary as it is still in common use in technical development today, albeit complemented by agile methods [59]. It is meaningful to specify the EEA as early as possible due to its interdependence with other domains and most electronic components [27]. When designing the EEA, developers need to evaluate these effects, because a large number of components and their multidisciplinary interfaces lead to high complexity. While it is already hard to design packaging and EEA loosely coupled by some requirements, the use of zonal EEAs further complicates this development procedure [27]. A cooperative design of zonal EEA enables full utilization of the benefits, resulting in a complexity and cost reduction of architecture and wiring harness.

Therefore the method described in this thesis is designed for the left branch of the V-model, supporting the iterative design process of the system. The introduced method aims to support the evaluation on the system level by analyzing the corresponding variants. Furthermore it is also beneficial on the component level to identify the influence of design changes. The evaluation of EEAs using this method is possible even with few information, leading to any early identification of meaningful variants. As more and more details become clear along the left branch of the V-model, which results in an updated knowledge, previous solutions lose their validity. As the method is designed to support the entire decomposition and implementation phase, it is beneficial to use the method again whenever important parameters have changes such as environment conditions, stakeholders or system parameters. Consequently, the featured procedure is applicable in the entire decomposition phase of the V-model, being most beneficial in the phases of system architecture design and component specification as depicted in Figure 2.4. In particular, in early phases of the development process when conceptual design decisions are made despite significant uncertainties, knowledge of architecture interdependence and parameters are useful.

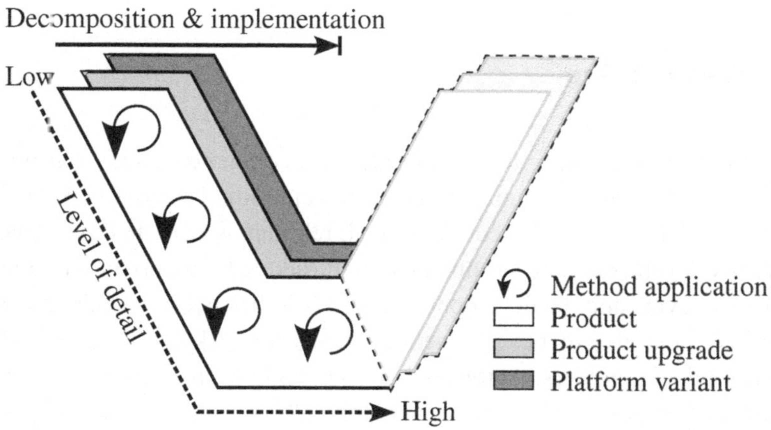

Figure 2.4: Scope of application for the proposed method

2.3.3 Tradeoff in Electric/Electronic Architecture Design

Constraints, evaluation criteria and trends form a solution space that is subject to a tradeoff. On the one hand, customers demand for new functionalities such as automated driving and connectivity, while on the other hand, they and legislators expect vehicles to be more efficient in the future and manufacturers strive to reduce the complexity of vehicles. Cost, quality and time further limit the options and strengthen the design conflict. As most architectures are part of a vehicle platform, they are adapted and reused for other vehicles with different requirements. Additionally, architectures need to be more flexible in the future, as manufacturers will update and upgrade their vehicles over the entire product life cycle and add new functions [60, 61]. To tackle these challenges new processes, technologies and tools are required. Two major technological advancements are the prevail of electric drives and automotive Ethernet. Today, it is certain that high-speed Ethernet will be used for the data backbone and low-speed Ethernet may partly substitute the established bus technologies [62]. Beside those, further progress, especially by means of processes and tools, is required. They need to be more flexible and adapt faster to a changing environment, especially with respect to software development, agile method will complement the procedure.

2.4 Research Gap

The interrelation of objectives and properties is often non-linear, causing complex interplay between them. Therefore, the optimum is a compromise, which has the overall best abilities considering all relevant aspects. In this chapter, the system's development environment was introduced, comprising architecture concepts, layers and the development process. It has been shown that not only the system but also its development is a complex task due to diverse design criteria and stakeholders. Automated driving, digitalization and electromobility are trends that pose challenges to the automotive industry as a whole, but especially to the EEA architecture. The requirements resulting from these trends are:

- Requirement 1: New connections respectively cable types due to high voltage power level
- Requirement 2: More connections due to additional electronic components
- Requirement 3: Tighter coupling of hardware and software components to realize new functions
- Requirement 4: Rising development speed due to a higher share of software components

Based on these requirements, the following challenges for the system and process are formulated:

- System challenge 1: Increasing system complexity due to a rising number of electronic components and software functions
- System challenge 2: Increasing wiring harness complexity due to new voltage levels, components and vehicle concepts
- Process challenge 1: Increasing integration effort due to stronger coupling of functions and systems
- Process challenge 2: System development under higher uncertainties due to faster software-driven innovations

One promising development direction is zone-centralization, for which geometric relationships are fundamental. As of now, there is no approach supporting engineers in the design of zonal architectures or in the transformation from domain- to zone-centralization. One reason for the lack of such a method is the interdisciplinary characteristic of the issue. On the one hand, knowledge of mechanical engineering for the packaging design, on the other hand, insights into electrical and network engineering for the design of architectures and wiring harnesses are essential. Additionally, a background in automotive engineering is helpful to formulate application-specific boundaries and parameters. To tackle the above mentioned challenges, zonal EEAs are a promising research field. The following research gaps are in the focus of this thesis:

- Research gap 1: Identification of metrics to evaluate the complexity of vehicle wiring harnesses

- Research gap 2: Simplification of the integration process due to a structured design method
- Research gap 3: Integration of existing approaches for other automotive EEA concepts for zonal EEA design
- Research gap 4: Simplification of the vehicle wiring harness with respect to the zonal EEA

In the following chapter crucial aspects are presented of zonal EEAs and their individual optimization. A holistic view is required for the optimization of the zonal design, therefore in Chapter 4 these individual approaches are combined to a holistic method. This includes the determination of ZCU positions, the zone boundaries as well as the design and positioning of electrical components, comprising wiring harnesses and the traction battery. By using an approach that combines these aspects and considers the constraints, the optimal zonal architecture is determined using the degrees of freedom in mechanical, electrical and system design. The aim of this process and the obtained results is to increase development efficiency by a demand-oriented design as well as a reduction in complexity of the entire system due to the simplification of components and interfaces. Furthermore, the identification of an optimal zone architecture provides the basis for modularity of many hardware and software solutions in vehicles.

3 Optimization of Electric/Electronic Architectures

EEAs are seen as vehicles' backbones and optimizing them often results in benefits for other systems. Although optimization brings progress for itself and other systems, it is hard to identify optima for EEAs due to its interdisciplinary and conflicting design criteria. Before the actions taken for this research are presented, it is necessary to point out the results of the previous action planning. In this chapter the problem is structured into core aspects. Based on these, suitable approaches to tackle them are identified. At first the criteria, constraints and the resulting tradeoff in EEA design are introduced to highlight the area of conflict and the complexity of the development process. In the following section the most influential hardware aspects such as packaging, traction battery, wiring harness, low voltage components as well as corresponding modeling and optimization approaches are presented.

3.1 Evaluation of Electric/Electronic Architectures

Modern product development needs to be user-centralized to ensure an efficient process and to fulfill the user's demands as advocated by the V-Model. As EEAs themselves do not directly interact with users, the requirements need to be transferred from the product level or from systems with user-interaction, such as multimedia or powertrain. In addition to the user, there are further stakeholders such as interacting systems that have demands on the EEA. Various authors have already defined criteria for a EEA evaluation, but the emergence of zonal architectures and a rising importance of the wiring harness requires an extension of those criteria [28, 45, 63, 64]. Consequently, this subchapter aims to describe new metrics as well as constraints for the evaluation of zonal EEAs.

J. Maier, *Optimization of Zonal Electric/Electronic Architectures Using a Coupled Modeling Approach of Package and Wiring Harness*,
Wissenschaftliche Reihe Fahrzeugtechnik Universität Stuttgart,
https://doi.org/10.1007/978-3-658-48188-9_3

3.1.1 Metrics

Metrics are used to measure properties or the behavior of a system and represent them in a concentrative and objective way [65]. The metrics presented in the following aim to provide a framework for consistent evaluation of architecture variants with respect to specific requirements [66]. While it is essential to assess the system itself, it is additionally reasonable to evaluate the wiring harness as its structure is most affected by the shift to zone-oriented architecture. Each metrics has their own dimension and therefore can not simply be added to a common complexity value. Nevertheless does each of them support the quantification of the complexity of the system as they follow the principal of homomorphism [67]. This means, that characteristics of a system change when complexity varies and hence the evaluation of the characteristics allows conclusions to be drawn about the complexity [68]. Starting with core properties such as cost and mass, which are crucial for the variant selection of both, continuing with functional criteria that are only meaningful on the architectural level and finishing with structural evaluation, which is more relevant for wiring harnesses.

The cost of a solution is often decisive concerning its realization in a product as it is desirable to satisfy the user's expectations at minimal cost. On the one hand this factor is essential for the manufacturer, as it is directly coupled to the profit, on the other hand the production cost has a large influence on the total cost of ownership. The total cost of ownership is a crucial factor for companies as it determines the price for their mobility service, however its determination is not meaningful without detailed knowledge of the use cases [28]. For individuals the cost and range of an EV are the most important aspects for the buying decision [69]. Although cost is a decisive factor for companies and individuals, its determination is hard without detailed information about the development, production and usage of the vehicle. Therefore, costs are considered in the evaluation, but they are limited to material or production cost, as the available data and cost models can only provide reliable information for those. The cost of an architecture C_{ee} is determined as the sum of the cost of components given by

$$C_{ee} = C_{lv} + C_{bat} + C_{wh} + C_{zcu} \quad , \qquad \text{Eq. 3.1}$$

wherein the costs are denoted by C_{lv} as low voltage components, C_{bat} as traction battery, C_{wh} as wiring harness and C_{zcu} as ZCU.

Packaging is the arrangement of all components in the vehicle [56]. Once the external geometry of the vehicle has been determined, components and interior are positioned. First, the passenger seats are determined and, based on this, the ergonomically relevant dimensions such as doors and steering wheel are defined [70]. In the next step, the position of the drive units is specified, followed by the remaining components, which also include those of the EEA architecture. Packaging must be defined as early as possible in the development process, since changes have far-reaching consequences for many components in the vehicle [71]. The cost of modifications increases exponentially during the development process, thus early identification of the final packaging is desirable [72]. Component mass and volume are decisive factors in the packaging development process, as both are limited and their distribution is restricted by several constraints. It is essential to minimize mass and volume to ensure an optimal positioning of the components in vehicles; furthermore, the mass is directly related to its efficiency and thus to its range. The mass of individual components can be determined by the number or size while it is hard to find a meaningful measurement for the volume. Furthermore, the effective volume of a wiring harness does not comprise its characteristics, whereas a measurement is necessary considering the distribution of the wiring harness. Consequently, the criteria length expansion $l_{e,wh}$ is introduced, characterizing the total length of all individual paths l_u in vehicles occupied by one or more cables. Together with the total length of all wires l_{wh}, it enables a meaningful assessment of wiring harness volume demand. They are calculated with respect to

$$l_{e,wh} = \sum_{u \in \mathcal{U}: u \geq 1} l_u \quad \text{and} \qquad \text{Eq. 3.2}$$

$$l_{wh} = \sum_{e \in \mathcal{P}} l_{c,p} + \sum_{d \in \mathcal{D}} l_{c,d} \quad , \qquad \text{Eq. 3.3}$$

wherein $\mathcal{D}$, $\mathcal{P}$ and $\mathcal{U}$ are the set of data and power cables as well as graph edges, while u, e and d are their corresponding elements. The length of each cable is indicated by l_c and the dedicated index.

Electrical balance, flexibility and State of function (SOF) are used to evaluate the functional behavior of the architecture. Electrical balance B_{ee} describes how much the amount of electric energy and power of each zone differ from the average. Flexibility defines the sensibility of the architecture to hardware changes [73]. SOF quantifies the safety behavior in case of a zone breakdown. The speed of technical development, especially of software, has increased in the last years, leading to users demanding for an ongoing development of functions even after product launch [10]. Upgradeability brings a lot of benefits, but there are two downsides that need to be mentioned. The first is the increase in variant number, the second is the demand for modularity to utilize an uniform upgrade process [24]. Both can be tackled by balanced zones with high flexibility, which are calculated using

$$B_{ee} = 1 - \frac{\sigma_{b,el}}{\mu_{b,el}} \quad \text{and} \qquad \text{Eq. 3.4}$$

$$F_{ee} = \frac{n_{f,2}}{n_{f,1}} \quad , \qquad \text{Eq. 3.5}$$

where the following variables are utilized: $\sigma_{b,el}$ and $\mu_{b,el}$ are the standard deviation and mean value of the electrical balance, which are both based on the zone dedicated battery cells. $n_{f,2}$ is the number of components that are assigned to the same zone after a hardware change and $n_{f,1}$ is the number of remaining electronic components. The corresponding simulation study for the determination of the flexibility is described in Appendix C.6.

The significance of safety in automotive systems raised by the advent of automated driving and electromobility, where both reduce the influence of the driver on the vehicle's behavior [74]. Since the functional architecture and the mapping on hardware is not the main focus of this thesis, it is not meaningful to conduct an analysis of the functional safety. However, EEAs are fundamental for realization of the functional architecture, hence it is necessary to consider safety in the design process [71, 75]. In a zonal architecture it is possible to enable zone separation in the case of an error, thus a worst-case scenario is a failure leading to a zone breakdown. This scenario is assessed by the rate of unaffected to total number of functions. SOF characterizes the component ratio that can still operate when the most powerful zone, by means of electric energy and power, shut down. It is thereby proportional to the number of available functionalities in the case of a safety critical failure in the power supply or data transmission, such as ZCU shutdown. The SOF of an EEA is determined by:

$$SOF_{ee} = \frac{1}{2} \cdot \left(\frac{E_{el,tot} - E_{el,max}}{E_{el,tot}} - \frac{P_{el,tot} - P_{el,max}}{P_{el,tot}} \right) \quad . \qquad \text{Eq. 3.6}$$

where $E_{el,tot}$, $E_{el,max}$, $P_{el,tot}$ and $P_{el,max}$ denote the total and maximum electric energy and power of an architecture.

Complexity is a rather nebulous term, a fact due to several reasons. One is the missing direct physical representation of complexity, another one is the inhomogeneity of systems. To be precise, it is dependent on many individual components and their interactions in a system comprising hardware and software. A third reason is the fact, that widely varying definitions of complexity have emerged in the different disciplines of engineering and science. Consequently, it is not possible in the course of this thesis to give a definition that satisfies the aspects of all disciplines. Nevertheless, complexity is a crucial measure to describe the effort needed for development, maintenance, upgrades and adaption to other variants [71]. While many components and aspects are considered by Riesner, the specific issues of wiring harnesses are neglected [76]. Maier has made assumptions regarding these parameters and applied the approach to zonal EEA, but the variety of approximated parameters makes comparisons hard [35]. Thus the complexity of wiring harnesses is determined using Sinha's approach according to [77]

$$c_{wh} = c_1 + c_2 c_3 = \sum_{i=1}^{n} \alpha_i + \left(\sum_{i=1}^{n} \sum_{j=1}^{n} \beta_{i,j} A_{i,j} \right) \frac{E(A)}{n} \quad , \qquad \text{Eq. 3.7}$$

wherein c_1 is the sum of the components complexity, c_2 is the complexity of the pairwise interaction and c_3 is the complexity of the interface arrangement. These three terms are determined using the complexity of the individual components α_i, the complexity factor of the interfaces $\beta_{i,j}$, the matrix indicating the interfaces between the components $A_{i,j}$, also known as adjacency matrix, the sum of the singular values $E(A)$ and the number of system components n. The singular values can be understood as a coordinate-by-coordinate scaling and therefore indicate the value distribution in the matrix.

In zonal EEA each zone has its own wiring harness leading to separate complexity values c_{wh}. The complexity balance B_{cmp} of a wiring harness quantifies the differences between these zonal wiring harnesses, which allow consideration of a balanced system design not only in terms of architecture, but also on the wiring harness level. Complexity balance is specified using

$$B_{cmp} = \frac{\sigma_{cmp}}{\mu_{cmp}} \quad , \qquad \text{Eq. 3.8}$$

wherein the standard deviation σ_{cmp} and mean complexity μ_{cmp} of all zones are used to determine the complexity balance.

By analyzing a system it is possible to identify that despite Eq. 3.7 embodying an universal metric, it may not sufficiently represent the entire scope of complexity. Development and production complexity are not considered within this metric, consequently the physical metrics need to represent them. Thus it is assumed, following the principle of homomorphism, that an increase in complexity balance, length, length expansion or mass results in an increased total effort and thereby represents development and production complexity. According to Wei, the division into zonal wiring harnesses does not necessarily reduce the complexity represented by a score, but it decreases the number of parts and variants and thereby simplifies the system and most associated tasks [78].

3.1.2 Constraints

In addition to metrics for measuring the quality of a variant, it is meaningful to limit the solution space using constraints to ensure the suitability of the results. The first constraint is the vehicle range, which is besides cost the most important aspect for individuals as well as companies when deciding for an EV [69]. The range is mostly determined by the battery capacity, the vehicle shape and its weight. Therefore the battery cell number is selected as a lower boundary with respect to the vehicle energy consumption in a Worldwide Harmonized Light Vehicles Test Procedure (WLTP) driving cycle.

The second constraint is an upper boundary to the zone quantity. Theoretically any zone number is possible, whereas from a practical point of view there are reasons for a limitation. It is not reasonable to have a higher number of zones than electrical loads, thus the ratio of electronic components to zones needs to be limited. Furthermore, the purpose of zones is to bundle data and power distribution of a geometrical share of the vehicle, hence the number of electrical components should not only be equal, but a multiple of the zone number to legitimate the centralization effort. Consequently, the zone quantity is limited to eight, with respect to existing numbers in research and development, forming an upper boundary. The identification of a suitable zone quantity is discussed in more detail in Chapter 3.2.

A third boundary condition is assigned to the aspect of safety. Ensuring safe operation of vehicles is even more critical for automated driving functions, thus a redundant power supply is necessary. Therefore, safety-critical components are connected to redundant power sources supplying them in case there is a malfunction of the power supply. The constraint is modeled by LV batteries and Supercapacitor (SC), which are connected by individual wires to the critical components.

The last constraint is the avoidance of package conflicts to ensure compatibility within the solution. A voxel based representation of the vehicle's geometry is used, enabling each component to occupy the corresponding voxels of its position and dimensions. The placement order is chosen by expected influence on the architecture ensuring suitable positions within the available space.

3.2 Vehicle Packaging in Zonal Architectures

The following section is focused on the arrangement of components and the division of the available packaging space into zones. Mass and volume play a key role in the context of packaging, thus they are determined in component specific models and introduced in the corresponding section. Table 3.1 contains an overview of existing methods for package modeling. PT and HVB are considered in all approaches as they comprise the components with the greatest influence on the packaging due to their large weight and volume. In addition there are further components such as electric loads, power supply and wiring harness that are noteworthy; however, these are either neglected or simplified in most methods. Many electrical loads have a comparatively low mass and small volume in relation to the whole vehicle [79, 80]. In particular, when considering the complexity of EEAs, neglecting its components leads to limitations in the expressiveness and usability of the solution due to their functional context, inhomogeneity and large number. Additionally, when conducting an efficiency analysis, considering the heating, ventilation and air conditioning system is inevitable [81]. It becomes apparent that the existing methods model the package, but mostly limited to the powertrain related components. Electric components are simplified e. g. the wiring harness is modeled by a regression based weight model. Consequently, the existing approaches are not suitable to model the crucial packaging aspects of EEA necessary to optimize the zonal architecture with respect to the wiring harness.

Table 3.1: Packaging optimization approaches

	Vehicle			Components						Packaging	
Reference	Passenger car	Commercial veh.	Longitudinal dy.	PT	HVB	HV loads	DC/DC converter	LV loads	Wiring harness	Volume model	Geometric model
[80]	x		x	x	x	x	x	x	(x)	(x)	x
[81]		x	x	x	x	(x)		(x)		(x)	x
[82]	x		x	x	x	x	x	x	(x)		(x)
[83]	x		x	x	x					(x)	
[84]	x		x		x			x		(x)	
[85]		x		(x)	(x)	(x)	(x)	(x)		x	
[86]		x		(x)	(x)	(x)	(x)	(x)		x	

x : Fully considered
(x) : Considered to a minor degree

3.2.1 Zone Design

In order to divide a vehicle into zones, a definition of the term *zone* is required, as well as criteria according to which the division is conducted. From the current state of the art (cf. references from Table 3.2) the following definition results:

Definition 3. *A **zone** denotes an aggregation of electrical and electronic devices in a vehicle. These are connected energetically and information-technically by a central control unit, which is designated as a zone control unit. The boundaries of zones are defined using spatial coordinates in the vertical, longitudinal and lateral directions of the vehicle.*

Table 3.2 contains an overview of publications dealing with zonal architectures and specifying zone numbers. The quantity of zones varies between two and eleven, with the majority of publications assuming four or six zones. The determination criteria of the zone quantity are not given in any of the mentioned publications. This is evident with respect to the varying and vague definitions of criteria and dependencies, some of them are described hereinafter. Bornemann declares that the number of zones varies depending on the requirements and complexity of the vehicle [87]. UNICARagil vehicles have four wheel drives and sensor modules, so Goth and the UNICARagil consortium designed a four-zone architecture [34]. Infineon has backed a number of four zones by pointing to the reduction in cable length and high availability of functions in this approach [88]. Schäfer describes the four corners of the vehicle as ideal locations for ZCUs, as this is where the data volume of the sensors as well as the energy demand of the actuators is highest [31]. Klaus-Wagenbrenner names the trade-off between wiring harness and ZCU costs as an indicator for the number of zones [89]. As a conclusion to these assumptions, a clear set of criteria and constraints is missing as well as their dependence on the zone quantity. Another essential term that should be delimited is topology, which is in this work understood as the arrangement of ECUs and the data lines connecting them, resulting in a network structure of computing nodes and communication paths. It should not be mixed up with the mathematical meaning, which focuses on the properties of geometric objects and their preservation under continuous deformation, or the general meaning of describing the geometrical characteristic of an object.

Zone design crucially affects the wiring harness characteristics. The wiring harness length is certainly dependent on the zone number and on the position of the ZCUs. As the zone number is the subject of this study, its variation is indispensable. In contrast the ZCU positions may be chosen independently, therefore it is meaningful to position them with minimal distances to the assigned electronic loads in order to reach the shortest wiring harness length possible. The position to other ZCUs and the central processing unit can be neglected, because they are connected to each other only by a few cables. The relative position of the ZCUs to the energy storage devices may have an influence on voltage drop and power dissipation. While power dissipation is minimized using the HV level, the voltage drop needs to be checked. Due to the spatial consideration of the electrical power demand, the drive concept has an influence on the zone design, which is discussed in detail in Chapter 5.3. The communication network is also taken into account by determining the cable cost of the data transmission lines. To summarize this aspect of the problem, the distributed components of a vehicle need to be grouped and the distance in between a group should be minimized.

Table 3.2: Common zone numbers for zonal E/E architectures concepts

Reference	[87]	[90]	[91]	[92]	[34]	[88]	[93]	[94]	[31]	[95]	[89]	[96]	[24]
Zone Number	2-4	2+3	4	4	4	4	4	4	4-6	6	6-11	8	10

3.2.2 Cluster Analysis for Zone Design

Cluster analysis is a discipline of data mining that aims to identify similarities in data and group them based on those. Therefore, it is a suitable approach for the zone design problem described in the previous section. As in cluster analysis the groups are not known beforehand and emerge during the process, it is referred to as unsupervised learning. The following sections provide an overview of clustering methods. There are representative, hierarchical, density-based as well as distribution-based methods. These are divided into hard-assignment and soft-assignment methods. Hard-assignment means that the data points can only be allocated to one cluster, in soft-assignment it is possible to classify the data into several clusters [97]. The following categorization is not complete, but should give an overview of different approaches for data clustering.

Representative-based clustering groups data $\tilde{x}_1, \tilde{x}_2, ..., \tilde{x}_n$ of a dataset $\mathcal{X}$ into a predefined number of groups k, which are called clusters or sets $\mathcal{S} = S_1, S_2, ..., S_k$. The database is described by n points in a d-dimensional space. Each of these clusters S_i has a centroid or center, which corresponds to the mean μ_i defined as

$$\mu_i = \frac{1}{\sum_{\tilde{x}_j \in S_i}} x_j \quad , \qquad \text{Eq. 3.9}$$

where i is the index of the clusters and x_j are the objects in S_i. In principle, it would be possible to compute all combinations to find the optimal clustering, but this procedure is practically not realizable. The number of possibilities increases accordingly $O(k^n/k!)$ and thus also the computational effort [98]. To achieve a meaningful solution with fewer computational operations, there are several methods, one of which is the k-means algorithm.

Hierarchical clustering distinguishes between divisive and agglomerative methods. The divisive methods are initialized with a cluster containing all objects. The clusters are divided with each step until finally one cluster per object remains. Agglomerative methods proceed in reverse, therefore each object is initialized representing a separate cluster. These clusters are merged in the process. The advantage of hierarchical approaches is the small number of parameters and the formation of hierarchical relations [99]. The downside is the high number of solutions, the overall complexity of $O(n^3)$ and thus high computational cost. Since no final partitioning of the data is created, but different solutions are available on all hierarchical levels, a post-processing is necessary for the evaluation and selection of the most suitable variant.

Density-based clustering partitions the data based on the density of the objects. Two widely used techniques are Density-Based Spatial Clustering of Applications with Noise (DBSCAN) and Ordering Points to Identify the Clustering Structure (OPTICS). DBSCAN classifies points as core, boundary, and noise points and requires two parameters: ϵ specifies the maximum distance allowed between two points for them to be considered concatenated and $minPts$ specifies the minimum number of chained points required for a cluster [100]. All points which are not within clusters are output separately as noise points. OPTICS uses the same parameters, but the procedure is modified to detect different densities [101].

Distribution-based clustering partitions the objects according to their distribution, e.g. Gaussian or normal distribution. The method assumes that each cluster has mean and variance. The advantage is that outliers are taken into account and non-linear relationships are considered, while the disadvantage is that the procedure can only be used if the distribution of the data is known [102].

An attempt has been made in this section to provide a brief summary of different clustering approaches and exemplary methods. Distribution-based procedures can not be applied as the distribution is not known. Although hierarchical clustering can be implemented, it is not efficient as the computational effort is large and additional post-processing is required. Density-based algorithms are not applicable, because of the varying densities of the data points and the fact that the remaining unsorted data points are noise points. The varying density makes it difficult to select an appropriate ϵ, as a high value could result in one cluster and a low in many noise points. Data points that are not grouped in a cluster are prohibited, as all components need to be assigned to a zone, respectively ZCU. Consequently a method of the representative-based clustering is used, namely k-means clustering, as it fulfills all requirements specified for the sorting procedure. The requirements are discussed in detail in Chapter 4.2.

3.3 Wiring Harnesses in Vehicles

A wiring harness comprises all cables for data and energy transmission connecting the electronic components in a vehicle, therefore it is often referred to as the vehicle's backbone. Due to the high number of interfaces and their large spatial distribution, it is one of the most complex components in modern vehicle development. Since the length and mass of a wiring harness are dependent on equipment, production year and vehicle type, various numbers exist quantifying its length and mass. Lengths of up to five km and masses of 60 kg to 80 kg are common in vehicles today [89, 96, 103]. The wide variety of vehicle models and their equipment causes a large number of variations in wiring harnesses, resulting in individual production processes with the necessity for manual work [104, 105]. All in all this results in the wiring harness having the 3rd highest cost and mass in cars today [106]. In-vehicle energy transmission is mainly realized by copper wires. Although aluminum brings some benefits, there are downsides limiting its use to rigid connections with high diameter, such as battery cables. Due to the increasing data traffic, automotive Ethernet will become necessary in the future for many applications and will replace the current technologies gradually [107]. The entirety of cables for power and data transmission is in the following referred to as the wiring harness.

3.3.1 Design

In the further course of this work the focus is put on the cable design and routing. The plugs cost are computed proportionately to the cable cost and their position for the components is considered in the routing. Figure 3.1 a) shows the wiring harness of a sedan, which is distributed throughout the vehicle and has typical tree-like branches. Figure 3.1 b) illustrates a model of the wiring harness with consideration of cables, connectors, paths and the bundling along them. Figure 3.1 c) visualizes the structure of cables used to transmit electrical data and power. Power cables used in vehicles often consist of individual strands, which are twisted and sheathed with an insulating material. The use of multiple strands increases the mechanical flexibility of the cables, which is required for complex installation paths and more robust against vibration. The design of data cables is dependent on the protocol or bus system and therefore is diverse. As this method aims to support any transmission protocol and technology, a worst case assumption is made, and thus low- and high-speed Ethernet is used, which has for all applications an equal or higher data rate than established systems. Accordingly the data cable visualized in Figure 3.1 c) is an Ethernet cable having four individual lines, each comprised of several twisted pair strands. To prevent crosstalk, each of them is shielded. Four pairs are combined to form an Ethernet cable and protected against electromagnetic effects from the outside with a common shield. An insulation layer provides electrical isolation and prevents mechanical damage. There are applications in the consumer sector using Ethernet for power transmission (Power-over-Ethernet), however this technology is under research and rarely used in current series-production vehicles [108]. In order to reduce the complexity of wiring harnesses, modular concepts are used nowadays, which are spatially limited and therefore have a

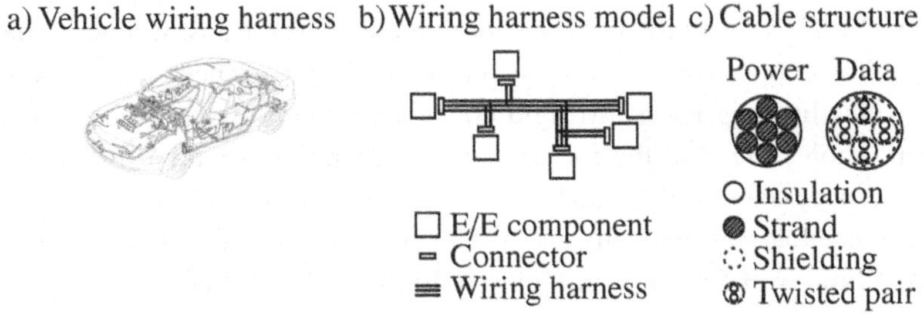

Figure 3.1: a) Vehicle wiring harness in a sedan [109], b) wiring harness model and c) data and power cable design

significantly lower number of variants [110, 111]. According to Pischinger, this reduces the number of variants to the sum of the variants of all harness modules, which allows the multiplication of the variants of all harness modules to be created. The aim is to cut costs by reducing the number of variants and increasing the quantity. For instance the part of the wiring harness within a door is designed for all options due to a low number of variants. The modules comprising the vehicle interior are manufactured to be customer-specific in order to cover the high number of variants at optimal cost. In addition to the modularization of wiring harnesses, optimal cable routing facilitates a reduction of weight, length and number of branches. To identify the minimum length and hence the least weight, the cable routing is considered as a SSSP problem in the following.

3.3.2 Modeling

While there is limited knowledge about the procedure for designing an optimal wiring harness, a number of factors is known that affect its design. A few of these factors are the high number of separate components, their distribution within vehicles, the tree-like structure and the high number of interfaces. Therefore, there is an inevitable need to limit the modeling scope to the crucial aspects. Since the aim is to evaluate zonal EEAs, it is necessary to model the cable paths and the electrical design to derive characteristics such as mass. The electrical properties of the power cables are reproduced by substituting the cables by electric resistors [112]. The aim of geometrical modeling is to determine the length and weight of the wiring harness as accurately as possible. On a theoretical basis linear regression models using empiric data are applicable, but there are two major issues. The first is that many properties are highly dependent on the cabin layout, cable routing and vehicle type, therefore deduction of characteristics is very specific and requires comprehensive models. The second is the lack of available data on wiring harnesses in mass-production vehicles. Reasons for this may be the minor role of the wiring harness for vehicle development in the past and difficulties quantifying its properties. One aspect introducing further complexity is the modeling of various cable types for data, for instance low- and high speed Ethernet, and power, such as low and high voltage. Diebig simplifies the calculation of cost and mass of cables by neglecting the different types. The mass is calculated with respect to volume and density of the conductor, while the cost is determined by multiplying the mass with the copper market price. Plugs and insulation are each considered through an individual factor [113]. The neglect of cable types is appropriate for the thermal simulation of Diebig, but not for the evaluation of EEAs. Fuchs uses a regression function for the determination of the low voltage wiring harness mass, but the estimation of the high voltage part is nebulous and data cables are neglected [82]. The simplification of Fuchs is meaningful in the context of weight estimation of entire vehicle concepts, but is not sufficiently accurate with focus on EEA optimization. Braun introduces a method to optimize the EEA focusing on the power supply system [28]. The approach defines plausible cost for each wiring diameter, but insulation and data cables are not considered. The insulation causes additional cost, which has a larger impact on high voltage cables. Braun defines the communication architecture on a system level in order to achieve high modularity enabling an independent design of power supply systems in vehicles. This method is highly valuable for the evaluation of power supply systems and EEAs. The rising number of electronic components will further increase complexity, cost and mass of vehicles. Consequently, a more detailed simulation of vehicle wiring harnesses in combination with EEA evaluation is essential to find the most suitable architectures for future vehicles.

To increase accuracy, the model for the routing comprises a rectangular three dimensional grid with relatively fine discretization compared to existing approaches. The grid is transformed into a graph determined by the set of edges

$\mathcal{U} = u_1, u_2, ..., u_k$ and nodes or vertexes $\mathcal{V} = v_1, v_2, ..., v_n$. A path p is defined by an ordered pair of edges and vertexes $(\mathcal{U}_P, \mathcal{V}_P)$. The length of cables is essential for all further calculations, therefore its determination for an individual cable $l_{c,i}$ in an unlimited environment is equal to the Euclidean distance between the coordinates of the starting vertex $v_{s,i}$ and the target vertex $v_{t,i}$ of the path according to

$$l_{c,i} = \left\| v_{s,i} - v_{t,i} \right\|_2 \quad . \qquad \text{Eq. 3.10}$$

The mass of a cable $m_{c,i}$ is calculated depending on its type. Power cables are determined using their length, the cross-sectional area A_i and the density of copper ρ_{Cu}, while data cables utilize the length and an Ethernet dependent mass factor $f_{m,Eth}$, which is set to 0.3 kg/m [114]. For their modeling it is often meaningful to differentiate between data and power cables to indicate individual characteristics. Therefore, in case-by-case functions there are two terms used to simplify the equations:

- 'Data' for $\exists d \in \mathcal{D} : p_i = p_d$ and
- 'Power' for $\exists p \in \mathcal{P} : p_i = p_p$,

wherein p_d, p_p and p_i are the paths in the graph of the data, power and the inspected cable. The mass of cables is determined using

$$m_{c,i} = \begin{cases} l_{c,i} f_{m,Eth} & \text{for Data} \\ l_{c,i} A_i \rho_{Cu} & \text{for Power} \end{cases} \quad . \qquad \text{Eq. 3.11}$$

Optimization of wiring harnesses can be done on the basis of various criteria, but a relation to the resulting costs is necessary for many scenarios to compare variants. Herein the cost is modeled using the length of a cable multiplied by the cost per length in Euro (EUR)/m according to

$$c_{w,i} = \begin{cases} l_{c,i} f_{c,Eth} & \text{for Data} \\ l_{c,i} f_{c,pow} f_{c,vl} & \text{for Power} \end{cases} \quad , \qquad \text{Eq. 3.12}$$

wherein three different factors are used to adapt the costs. $f_{c,pow}$ or $f_{c,ETH}$ is the cost per length, while the voltage level of the cables is considered using a specific cost factor $f_{c,vl}$.

3.3.3 Optimization

Wiring harnesses are highly application specific and therefore are often not optimized with respect to their utilization [115]. The EEA development additionally poses some further challenges to ensure practicality. It is necessary to consider vehicle specific characteristics, packaging aspects and the integration of requirements from other systems or the entire vehicle. Table 3.3 shows different methods of wiring harnesses optimization for varying purposes. Some are universally usable, others are vehicle-specific, and one procedure originates from aircraft engineering. Most of these methods neglect the electronic components and focus exclusively on the routing of wiring harnesses. Braun instead considers the electronic components to optimize the routing paths with respect to their requirements [28]. Lin and Huang additionally consider the optimization of the bundling by modeling the wiring harness as a Steiner tree problem [116, 117]. The neglect of the wiring harness environment may result in a theoretically optimized path, but component- or vehicle-specific requirements are not considered. A different approach is taken by Chandra et al., who designed a wiring harness backbone for Tesla Inc. that follows both sides of the vehicle's longitudinal axis and is connected to a battery by a cross bar [118]. Components are connected to the backbone, which is used to transmit power and data in the vehicle. By using such a backbone, production could be automated, reducing assembly and manufacturing cost [119]. Modularity can be further improved by merging the concepts of zonal wiring harnesses, inter-zonal backbone and variant specific modules. Development can also benefit from this approach, as the backbone no longer needs to be adapted to each variants. Instead, it can be designed according to the variant with the highest power and communication requirements at little additional costs, but there are currently no known implementations of this concept. Consequently, the optimization problem is considered as the identification of a shortest path between a starting and an endpoint in a restricted environment.

Table 3.3: Wiring harness optimization approaches

	Vehicle			Components					Packaging		
Reference	Passenger car	Commercial veh.	Longitudinal dyn.	PT	HVB	HV loads	DC/DC converter	LV loads	Three-dim. routing	Requirement transfer	Cost
[28]	x		x	x	x	x	x	x	(x)		x
[50]	x								(x)	(x)	(x)
[113]	x				x	x	x	x			(x)
[120]									x	x	
[121]										(x)	x
[122]									x	x	x
[123]	x	x							x		
[124]		x							x	(x)	
[125]									x		x
[126]									x		
[116]									x	x	
[117]									x	x	

x : Fully considered
(x) : Considered to a minor degree

3.3.4 Shortest Path Problems

Since wiring harnesses in vehicles transmit power and data between source and sink, the routing of its cables corresponds to a SSSP problem. Identifying the shortest path between two locations occurs in a wide variety of engineering applications. One way to numerically solve such problems is to transform them into a graph of edges and nodes. The vehicle structure is modeled by a three-dimensional graph. The objects, in this case the start, junction and destination points correspond to the nodes, the connections between these objects to the edges. According to the edge properties, a distinction is made

between graphs. Undirected graphs allow edges to pass in both directions, while directed ones allow only one direction. Multigraphs permit several edges between two objects, planar graphs exclude the crossing of connections if the graph is represented on a plane. Edges can be weighted to create comparability between connections, thus the weights $\mathcal{W} = w_1, w_2, ..., w_k$ are defined for the edges. Consequently weighting of edges can result in a completely different optimal path p_{min} compared to the length optimized path. If a start and end node cannot be connected $p_{min} = \infty$ applies. The shortest path for a start node v_s and an end node v_t thus corresponds to
$\forall p_{min} : v_s, v_t \in p_{min} \wedge p_{min} < \infty$

$$p_{min} = \min\left(\sum_{j \in \mathcal{U}(p)} u_j w_j\right) \quad . \qquad \text{Eq. 3.13}$$

There are several algorithms for solving SSSP problems. Three common ones are breadth-first search, Dijkstra's algorithm and the A* algorithm. A breadth-first search begins at a starting node and checks if the neighboring nodes have already been visited or if it is the element being searched for [127]. If one of the nodes is the demanded element, the algorithm determines the shortest path to it and ends the search. If this is not the case, all non-considered nodes are visited in the following iteration steps and the search is continued at their neighboring nodes. The disadvantage of this method is the uniform propagation in all directions, leading to extended computing times.

Dijkstra's algorithm enables the calculation of the shortest path from a starting point to all points of a graph or to a target node [128]. If the destination node is already known, as in this use case, it is restricted to the shortest path between the start and the end nodes. Each node is characterized by the properties predecessor and distance from the start. Nodes are categorized into three lists for this purpose: Visited and unvisited nodes as well as a queue. At the beginning all elements belong to the list of unvisited points, except the starting point, which is the only one in the queue. Beginning from the start node, all neighboring elements are added to the queue and sorted by distance. The queue is processed according to the shortest distance from the start node while visited nodes are shifted to the corresponding list. The neighboring elements of the visited nodes are selected and taken over into the queue. Visited points are not inspected again, but if a shorter distance is determined for them, distance and predecessor are updated. By taking the edge cost into account, the path with the lowest cost is determined instead the one with the shortest distance.

The A* algorithm also uses three lists: Unknown nodes, known nodes (open list) and fully explored nodes (closed list). In addition, a value is assigned to each node to implement the heuristic, which is also referred to as the f-value [129]. Beginning at a starting node, the neighboring nodes are added to the open list. Depending on their f-value, the nodes from the open list are considered in ascending order. Completely examined nodes are moved to the closed list. Thus, the heuristic can be used to control the search direction of the algorithm. The runtime of the A* algorithm is the fastest, as long as the heuristic maps the actual relationships correctly. For the wiring harness routing there are no heuristics available. In addition, it should be said that due to the large number of components and stakeholders, it is not clear whether heuristics can be transferred to other variants or general assumptions can be formulated. Instead Dijkstra's algorithm is used as it facilitates the consideration of requirements by modification of the edge cost. Furthermore it is most efficient in an environment where no reliable heuristic data is available.

3.4 Traction Batteries in Vehicles

In the field of automotive engineering various terms are used for the battery providing energy on the high voltage level of the vehicle. While HVB is an electrical term based on the voltage level, traction battery is the name that has emerged in the automotive industry as its most important purpose is to supply the powertrain in EVs. The following chapter contains a description of the design and modeling of traction batteries as well as the optimization of it using the NSGA-II algorithm.

3.4.1 HV-Batteries

A HVB in vehicles comprises a large number of lithium-ion cells connected in parallel and series. Further components of the HVB are the Battery management system (BMS), for monitoring and diagnosing the battery cells, the parallel and serial interconnections as well as a contactor for disconnecting and connecting the battery to the vehicle electrical system [130, 131]. These components are embedded in a housing, although this could be replaced in whole or partly by a self-supporting structure in the coming years [132]. As the cells are the essential component of HVBs, it is reasonable to discuss their structure and properties in more detail. Electrodes and active materials are enclosed in housings, often made of metal, using one of the three most common cell designs, which are cylindrical, prismatic and pouch cells [133, 134]. In cylindrical and prismatic cells the electrode and separator layers are stacked, coiled and integrated into a solid housing. In contrast, the housing of pouch cells is flexible, because it consists of an aluminum composite foil enclosing the stack of individual cell layers. Each cell design has specific mechanical, thermal and electrical properties bringing different advantages and disadvantages. The selection of the most suitable design is therefore complex, as the optimum must be achieved with respect to its application, the cost and its characteristics. For the simulations in the course of this thesis, cylindrical cells were selected due to their overall good performance.

Mass and volume of HVBs are essential for the geometrical modeling. As the cost is one of the dominant criteria in the automotive industry besides technical properties, it is valuable to use it for comparison of different technical variants. The total battery cost $C_{bat,tot}$ is determined using the number of cells n_{bat}, the factors for the cost increase from package to system level $f_{bat,C2S}$ and the cost per battery cell $f_{c,cell}$. For the auxiliary components, such as BMS and connectors, $C_{bat,aux}$ is added according to Kries [135]. The total battery cost is determined using

$$C_{bat,tot} = f_{bat,C2S} f_{c,cell} n_{bat} + C_{bat,aux} \quad . \qquad \text{Eq. 3.14}$$

The HVB total mass $m_{bat,tot}$ and total volume $V_{bat,tot}$ are estimated by multiplying the number of cells n_{bat} by the package efficiency factor for mass

$\eta_{bat,m}$ or volume $\eta_{bat,V}$ and the mass m_{cell} or the volume V_{cell} of a single cell according to Eq. 3.15 and Eq. 3.16. The efficiency factors are taken from the work of Löbberding et al. and Nicoletti, whereas the cost, mass and volume of the cell are chosen with respect to market available cells. The number of cells is determined on the basis of the desired range [80, 136].

$$m_{bat,tot} = \eta_{Bat,m} m_{cell} n_{bat,m} \quad \text{and} \qquad \text{Eq. 3.15}$$

$$V_{bat,tot} = \eta_{Bat,V} V_{cell} n_{bat,V} \quad . \qquad \text{Eq. 3.16}$$

3.4.2 Battery Design

The traction battery is the component with the highest mass and volume share in today's EVs according to Appendix A.1. Small savings in the battery through optimal and demand-oriented design can thus lead to significant increases in efficiency and comfort. Therefore it is meaningful to develop a traction battery with respect to the vehicles energy consumption and required range. The energy consumption of a vehicle drive W_{drv} results from the sum of the resistance forces F_{res} multiplied by the integral of the time-dependent velocity profile $v_{veh}(t)$ in the interval from 0 to τ according to [28]:

$$W_{drv} = \int_0^\tau F_{res} v_{veh}(t)dt \quad , \qquad \text{Eq. 3.17}$$

$$F_{res} = F_d(v_{veh}(t)) + F_r + F_g + F_a v_{veh}(t) \quad . \qquad \text{Eq. 3.18}$$

The air resistance F_d, the rolling resistance F_r, the gradient resistance F_g and the acceleration resistance F_a are determined using [110]:

$$F_d = \frac{1}{2}\rho_a c_w A_{veh} v_c(t)^2 \quad , \qquad \text{Eq. 3.19}$$

$$F_r = g m_{veh} c_r \quad , \qquad \text{Eq. 3.20}$$

$$F_g = g \sin(\alpha) m_{veh} \quad \text{and} \qquad \text{Eq. 3.21}$$

$$F_a = (e_i m_{veh} + m_{load}) a_{veh}(t) \quad . \qquad \text{Eq. 3.22}$$

Therein ρ_a denotes the air density, c_w the drag coefficient, A_{veh} the projected frontal area, v_c the relative velocity, m_{veh} the vehicle mass, c_r the rolling resistance coefficient, α the slope angle of the road, e_i the mass factor of the powertrain inertia, m_{load} the payload and $a_{veh}(t)$ the time-dependent acceleration of the vehicle.

In addition to the energy demand of the propulsion system, it is crucial to consider the consumption of auxiliary components for an accurate battery design. These comprise components of the thermal management, including air conditioning, data processing, sensor technology for automated driving, as well as display and control elements. Different modeling depths are possible with respect to the available information. Dynamic load, event-based as well as average power demand are the most common ways to model the behavior of electronic components. According to this order, the accuracy decreases to the same extent as the necessary information [28, 137]. Measurements or electrical parameters are not available until the prototype is built and the dynamic load variations are relatively small compared to the absolute demand of the loads. Thus for each load, either an event-based state model is created or an average demand is determined. Since some of Braun's data is outdated, a new data set is created in the work for this thesis, considering the existing data as well as hardware for automated and connected driving functions, which can be found in Appendix A.2. Another data set of electric loads was designed based on the UNICARagil vehicles and is presented in Appendix A.3.

3.4.3 Genetic Algorithm for Multi-Criteria-Optimization

Since traction batteries provide most of the energy in EVs, optimizing the energy storage can decisively improve the vehicles properties. Optimization is performed with respect to various aspects, often focusing on battery capacity, packaging, voltage stability, thermal behavior or minimum cost. Voltage stability will not be investigated further in this thesis, as Braun identified by a design space exploration that all their EV variants meet the requirements regarding voltage stability [28]. Table 3.4 contains a list of various methods for optimizing traction batteries. Apart from the approach of Becker, all others references listed in the table simplify the vehicle environment to the driving resistance equation and neglect the EEA components, which can lead to unfeasible results as pointed out by Fuchs [13, 82]. The battery in his concepts is determined based on energy demand or range, but mass and volume of the battery in some of the concepts exceed the available values due to the low energy density at the beginning of traction battery development.

Table 3.4: Traction battery optimization approaches

	Vehicle			Components					Battery	
Reference	Passenger car	Commercial vehicle	Longitudinal dyn.	PT	HVB	HV loads	DC/DC converter	LV loads	Electrical model	Geometrical model
[13]	x		x	x	x		x	x	x	(x)
[14]	x		x		x				x	x
[138]	x		x		x				x	x
[139]	x		x		x				x	x
[140]	x				x				x	x

x : Fully considered
(x) : Considered to a minor degree

In many technical systems an improvement of one property is accompanied by a restriction or degradation of another, which is particularly true for the traction battery. The different properties vary depending on the cell type, but none of them surpasses another in all properties. As the number of criteria and the complexity of the relationships between them increase, it is more difficult to describe the problem analytically. Due to the increasing availability of computing power in the last decades, numerical methods such as multi-objective optimization are now commonly used. One problem with the use of numerical methods is the identification of local extremes. These are solutions in the target space which have a minimum or maximum value within a certain environment. If such a point is interpreted as a global optimum and the search is terminated, the solution found is most likely not to be the most suitable one. This is not necessarily a problem especially when there are many criteria or a large solution space, but it needs to be ensured that the solution found is good enough.

Stochastic methods can help to avoid this issue by using heuristics or random variables. Evolutionary or genetic algorithms are based on the logic of genetic reproduction of living organisms. For their application, first an initial population is generated, usually at random. Each individual of this population fulfills the criteria differently on the basis of its characteristics. The quality of the individuals is determined by means of a fitness function. Starting from this initialization a continuous procedure begins, which consists of recombination, mutation, evaluation and selection. In the recombination phase elements of the population are combined to generate new ones. Mutation forms the random component in the algorithm, as individual elements are casually mutated resulting in the generation of individuals that are significantly different from the previous generation. This enables escaping areas of local optima [141]. The previous generation as well as the newly created generation are evaluated using a fitness function and the best are selected as the new population. This continues until a termination criterion is reached such as a maximum number of iteration cycles or a target value for the solution [142]. For multi-objective problems, there is a large number of optimization methods. One that is highly versatile, achieves reasonable results, and shows good performance is the NSGA-II algorithm, which was selected for use within the work of this thesis [143]. The NSGA-II and the application to the problem is discussed in detail in Chapter 4.3.

3.5 Low-Voltage Powernet Components

This section presents the main hardware components of the low voltage powernet in vehicles and covers their essential functional aspects. Therefore, the corresponding components are introduced and cost, mass as well as volume models are formulated using linear regression. It is assumed that ZCUs will completely take over the tasks of switches, power distribution units as well as

Direct current/direct currents (DC/DCs) [89]. Consequently it is meaningful to model them and approximate the ZCU using those data.

3.5.1 Low-Voltage Batteries

LV batteries comprise several lead-acid battery cells which are connected in series, resulting in a nominal voltage of 12 V. In vehicles with internal combustion engines, it is used as an energy buffer during operation and as an energy source for the starting procedure. EVs still need a LV battery, because it serves as a redundant energy source in the event of failures of the HVB, buffers power fluctuations and provides energy for the starting process before the HVB is connected to the vehicle's electrical network. Lead-acid cells are mainly used yet, but it is assumed that lithium-ion battery cells will also be established in the coming years for LV batteries [144]. Although lead-acid batteries possess a different chemistry, several aspects discussed in the context of traction batteries apply to them as well. The cost $C_{bat,LV}$, mass $m_{bat,LV}$ and volume $V_{bat,LV}$ are approximated by

$$C_{bat,LV} = 0.8902\frac{EUR}{Ah}C_n + 47.595EUR \quad , \qquad \text{Eq. 3.23}$$

$$m_{bat,LV} = 0.2701\frac{kg}{Ah}C_n \quad \text{and} \qquad \text{Eq. 3.24}$$

$$V_{bat,LV} = 0.1067\frac{m^3}{Ah}t10^{-3}C_n + 1.6003m^3 10^{-3} \quad , \qquad \text{Eq. 3.25}$$

wherein C_n is the nominal battery capacity. The equations were determined using linear regression on a data set of freely available components. The data points used for the regression as well as the coefficients of determination R^2 are listed in Appendix B.3.

3.5.2 Supercapacitors

SCs are double-layer capacitors consisting of two electrodes, a separator and an electrolyte [145]. When a voltage is applied to SCs, charge separation enables the storage of electrical energy [146]. By connecting double-layer capacitors in series, the voltage of the SC is matched to the application. SCs differ from batteries in that their energy density is low and power density is high in comparison with batteries of similar mass and size. Due to the low energy density they are not suitable as the main energy storage, while the high power density enables fast charging and discharging. Thus SCs are most suitable for the smoothing of power peaks, which may occur intentionally for the case of recuperation or unintentionally due to a lack of power for the case of a battery failure. The cost, mass and volume of the SCs are also modeled by a linear regression based on a data set of available SCs according to

$$C_{SC} = 6.7953\frac{EUR}{Ah}E_{SC} + 824.55EUR \quad , \qquad \text{Eq. 3.26}$$

$$m_{SC} = 0.1447\frac{kg}{Ah}E_{SC} + 6.8542kg \quad , \qquad \text{Eq. 3.27}$$

$$V_{SC} = 0.1658\frac{m^3}{Ah}10^{-3}E_{SC} + 5.6716m^3 10^{-3} \quad , \qquad \text{Eq. 3.28}$$

wherein E_{SC} is the electric energy that can be saved in the SC.

3.5.3 DC/DC-Converter

DC/DC converters transform the electrical voltage, which enables energy to be transferred between two powernets with different voltage levels. The most powerful DC/DC converter in vehicles is the one connecting the two voltage levels of the power supply system, more specific the high-voltage side and the low-voltage level [110]. They are also used for other purposes in the vehicle, e.g. voltage stabilization within ECUs [112]. Voltage converters can be either unidirectional or bidirectional. Unidirectional converters are preferred especially in budget cars to save cost, while the bidirectional ones enable vehicle-to-grid applications and may bring additional safety. If the input is lower than the output voltage, bidirectional converters operate as boost converters, in the opposite case they are buck converters. Bidirectional voltage converters that are able to switch between these modes are denoted as boost-buck converters. The efficiency of DC/DCs for voltage levels typically used in vehicles is approximately between 85 % and 96 %, depending on load and voltage level [147, 148]. It is possible to model their behavior by equivalent circuit models, as the dynamics of the electrical network, as well as the exact efficiency is dependent on them. Since the dynamics within the networks are neglected for this work, a system model is sufficient according to

$$P_{el,vc,in} = P_{el,vc,ou} \eta_{el,vc} \quad , \qquad \text{Eq. 3.29}$$

with $P_{el,in}$ as input power, $P_{el,ou}$ as output power and η_{vc} as efficiency. Thus, for the buck converter case the current at the HV level $I_{vc,in}$ is given by

$$I_{vc,in} = \frac{U_{vc,ou} I_{vc,ou}}{U_{vc,in}} \quad . \qquad \text{Eq. 3.30}$$

It is not beneficial to describe the models for cost, mass and volume in detail as the function for converting the voltage is integrated in the ZCUs.

3.6 Delimitation of the Solution Space

The introduction of zonal EEAs poses challenges for various disciplines and architecture layers. It should be emphasized that this work is not able to handle all of them. Consequently, the scope is limited as described in Chapter 1 and Section 2.4. In this chapter, the disciplines and architecture layers that are in the focus of this thesis were discussed in more detail. These include the evaluation of EEAs, the vehicle packaging, as well as the design and integration of the low-voltage powernet components, the traction battery and the wiring harness.

The evaluation of EEAs, respectively their complexity, cannot be done with one single value. In fact, abstract complexity, physical and structural metrics are defined to take into account the multidimensional nature of the system. In addition, some constraints are defined to respect boundaries of the vehicles as a higher level system or technological limitations. In the packaging sections, the zonal distribution and its consideration in the packaging are discussed. While this is an essential question in the design and integration of zonal EEAs and in this work, there are only vague assumptions in the literature. Consequently, the problem of electronic component allocation to zones and identification of zone centroids is considered as a clustering problem and therefore, cluster analysis and the established methods are introduced. k-means clustering is chosen as the most promising approach due to its input and output parameters. To utilize the full potential of zonal EEAs, it is necessary to optimize the wiring harness. Therefore, a routing model is described based on a graph model. Furthermore, the shortest path problem and algorithms to solve the same are introduced. As Dijkstra's algorithm performs best without heuristics, it is selected to determine the shortest path between a starting node, such as a ZCU and a target node, which can be any electronic component. Although the traction battery is not a core aspect of the zonal EEA design, it is the component with the highest mass and volume in vehicles. Therefore, modeling approaches, constraints and algorithms to design a suitable battery under those conditions are described. The NSGA-II is a genetic algorithm and performs well for problems with two to three optimization targets. In addition, modeling approaches for LVs batteries, SC and DC/DC converters are described, as these should also be considered in the system design. In the following section, a more detailed definition of the zonal architecture and the application and adaption of the introduced methods to the problem are aimed for.

4 Design of Zonal Electric/Electronic Architectures

So far implementation of zonal architectures has been limited to prototype vehicles and systematic approaches for their design have not established. To tackle this issue, the individual solutions discussed in Chapter 3 with respect to the system and its environment described in Chapter 2 are joined to realize a holistic design method. Hence a method is introduced for designing a zonal architecture based on the optimization of traction battery, power distribution and wiring harness, while ensuring compatibility with vehicle packaging. Figure 4.1 presents the scope of the method in the context of an architectural definition workflow. The number of process steps and their interdependence punctuate the complex development environment. The aim of the method is to find a system wide approach considering many, but not all, of the displayed process steps. As there is often only a vague understanding and no common sense of zonal EEAs and corresponding terms, it is meaningful to define those terms first. In further sections the separate design and optimization steps as well as their integration in the methods are explained.

4.1 Zonal Architecture Definition

The terms zonal EEA, ZCU and zone are used more often now that an integration into series vehicles is getting closer. While there is a common understanding of zonal EEA, in contrast zone and ZCU are used with different meanings, resulting in a fuzzy understanding of these two. The aim of the following chapter is to tackle this issue by defining and delimiting both terms.

J. Maier, *Optimization of Zonal Electric/Electronic Architectures Using a Coupled Modeling Approach of Package and Wiring Harness*,
Wissenschaftliche Reihe Fahrzeugtechnik Universität Stuttgart,
https://doi.org/10.1007/978-3-658-48188-9_4

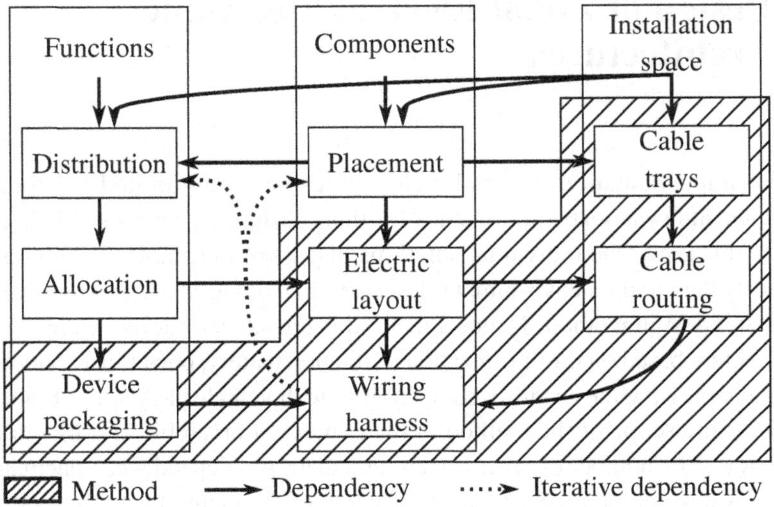

Figure 4.1: Workflow for the architectural definition and scope of the method adapted from [110]

4.1.1 Semi-formal Description

For the definition of an interdisciplinary model it is crucial to use the discipline independent modeling language SysML. Braun and Obergfell already defined SysML models for EEA, but in a more general way and not specialized on zonal architectures [149, 150]. Figure 4.2 shows the block definition diagram for a zonal EEA by Maier, which is referred to in the optimization process [151]. As the interconnectivity of infrastructure and vehicles is increasingly important, the EEA is put in terms of relation to the vehicle and the entire mobility system. A zonal EEA comprises zones and a central computing unit. Each zone includes one ZCU, which inherits the functionalities of a DC/DC converter, a PDU and a switch [152]. Consequently, each ZCU is assigned to both communication system and power supply. Furthermore, at least one electric component and, for the case that the zone should operate independently, one HVB cell belong to each zone. In addition to the assigned components there are shared ones, that are available for several zones e.g. the LV battery and the 48 V SC. A vehicle is separated in zones by separation into volume elements. The zonal architecture definition includes only few elements to reduce complexity and increase applicability in all aspects. A functional or software view is not defined as it is assumed that ECUs in the future will be more generic, especially the ZCUs, hence more function independent [153–155]. This will allow vehicle ECUs to shift functions between hardware. This system definition strives to create as few as possible limitations for the functional or software architecture; moreover, the ZCU should be considered as data and power hub in each zone. The hardware architecture in Figure 4.2 is designed with respect to several trends and present technologies such as the zonal module design of Texas Instruments or protocol and architecture concepts of Infineon Technologies [91, 156].

The role of the ZCU is not fully defined yet as several of its design aspects are not clear, especially its processing power is discussed in a wide range [157]. On the one hand it is referred to as a hub for data and power that concentrates on the forwarding of both, on the other hand it is seen as an ECU that additionally deals with the data processing within a zone. The specification of the processing power is not defined by this model to ensure its applicability with all characteristics between those two extremes. In the course of this dissertation it is understood as follows. Independent of the implementation, it is the central node of a zone and therefore necessary for each of them.

Definition 4. *A **ZCU** is an ECU that manages data and low voltage electric power transmission in a zone. This comprises the forwarding of both and optionally their compression or transmission. Therefore, its capabilities can range from pure assistance to full responsibility of the zone's components by functions such as data pre-processing and zone-individual energy management.*

The properties of the electrical components are crucial as the characteristics of the zone such as electric capability, data transmission and the spatial expanse are determined with respect to them. As redundant power supply is essential especially for autonomous driving, the HVB design should be incorporated in the zone development process. The mapping of battery modules to zones enables the independent operation of each zone, albeit such a design leads to restrictions in the choice of the battery size. It is still necessary to accommodate all modules in one housing to facilitate high energy density and good safety performance at a reasonable cost. The module size is determined by evaluating the zone with the lowest performance and the ratio to the other zones. In addition to the traction battery there are optional 48 V SCs and 12 V batteries, which are shared by several zones as backup power supplies for the corresponding voltage levels just in case a ZCU stops operating. The assignment of battery modules to zones enables a logical separation of the HV power supply and therefore an independent operation, but this is only a concept yet. However, Porsche already implemented a logical separation of its traction battery into two individual modules for charging in the new *Premium platform electric* [158].

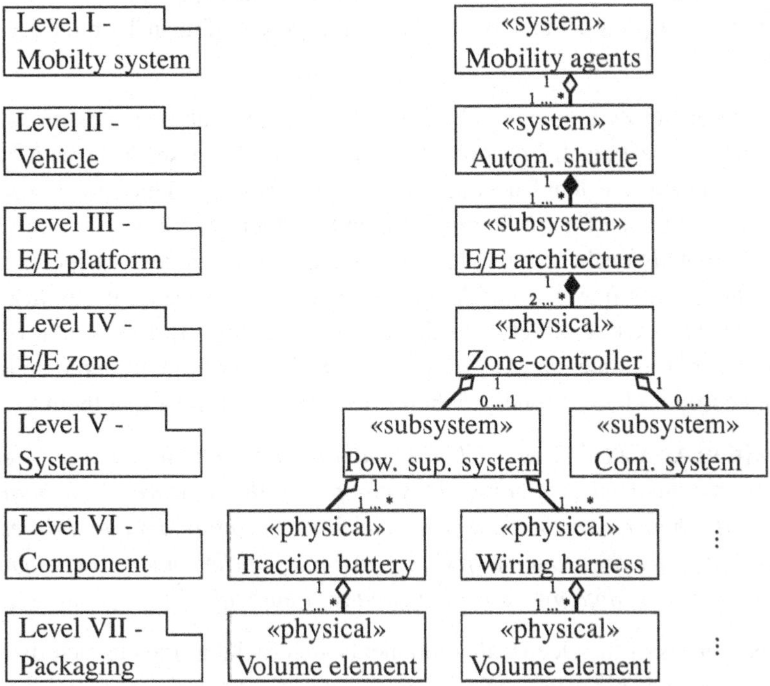

Figure 4.2: Zonal architecture definition using SysML adapted from [151]

4.1.2 Interdependence with Functional Architectures

The functional architecture is according to Chapter 2.2 the basis for the layers below, however zonal EEAs can support the decoupling of hardware and software. These two aspects seem to be opposing at first, but can be harmonized on closer examination. The method introduced in this dissertation is mostly software independent, justified by the fact that computation power in the automotive industry has strongly risen in the last few years, thereby supporting the generalization of ECUs. Generic ECUs have several advantages such as flexible reallocation of functions for updates or failure handling, which both are important in the context of software-defined vehicles and automated driving. Nevertheless, there are two drawbacks that go along with the assumption of generic ECUs as ZCUs. The first is an imbalance in communication effort and processing power caused by the different demand placed on both by functions. To tackle this issue the number of interfaces for all ZCUs is aligned. The second issue is that the size of the ZCU may vary depending on the processing power. Following the assumption of Burkacky the integration of functionality into ZCUs will decrease with rising SAE level, consequently this issue will solve itself [10].

4.2 Spatial Zone Optimization

The optimization of zonal EEAs comprises the structuring based on inhomogeneous electric components in a spatial environment with limitations. Therefore, it is necessary to add post-processing mechanism to the k-means algorithm, introduced in Chapter 4.3, in order to ensure compatibility with vehicle packaging and practicability.

4.2.1 k-means Clustering

The classification of electrical loads into the zonal architecture can be realized only by a few clustering methods. Since the components should belong exclusively to one zone and thus only to one ZCU, they must not be grouped into multiple clusters, therefore only hard-assignment procedures can be used. A similar design of the ZCUs leads to several benefits, such as a cost decrease due to rising quantity and easier integration because of the significant similarities in all zones. Therefore, all clusters shall have the same number of data points, respectively electrical loads, deviations are only allowed if they cannot be divided equally among the number of clusters. The number of zones should not depend exclusively on the distances of the components, since other criteria also play a role in the development of a EEA such as functional safety, transmission protocol or voltage level. Based on a literature review and evaluation of the technology level of ZCUs, the typical number of zones is determined as $\mathcal{Z} = 1, 2, ..., 8$, according to Chapter 4.2 and Table 3.2. Consequently, the method does not need to determine the zone quantity, instead it should identify the optimal allocation of data points to a cluster as well as the cluster centroid for a defined number of clusters. The centroid corresponds to the position of the ZCU. Based on these requirements the k-means algorithm is most suitable, as it groups data objects based on distance data in a hard assignment procedure and determines their centroids.

The k-means algorithm extends the representative-based clustering approach with a scoring function, often based on Sum of squared errors (SSE). At first k centroids are randomly generated in the search space, followed by the assignment of data points to their closest centroid forming the clusters S and a subsequent computation of the SSE according to

$$SSE(S) = \sum_{i=1}^{k} \sum_{j=1}^{n} \left\| \tilde{x}_j - \mu_i \right\|^2 \quad . \qquad \text{Eq. 4.1}$$

The objective is to identify the clustering with the minimum SSE, which is defined as

$$S^* = arg \min_{S} \{SSE(S)\} \quad . \qquad \text{Eq. 4.2}$$

After the evaluation using Eq. 4.1 the centroids positions μ_i^* are updated, corresponding to the mean value of all data points in the corresponding cluster, i.e.

$$\mu_i^* = \frac{1}{n} \sum \left\| \tilde{x}_j - \mu_i \right\|^2 \quad . \qquad \text{Eq. 4.3}$$

The assignment of the data points to clusters, the evaluation by means of SSE, as well as the updating of the centroids are performed iteratively until the termination criterion is reached. Once the centroids stop changing, it can be assumed that the algorithm has converged. It should be noted that the optimum reached may be a local optimum [98].

4.2.2 Initial Centroid Positions

There are only few parameters available to adjust the behavior of the k-means algorithm. One of these are the initial position of the clusters, which embodies the starting centroid and hence the assigned data points. These positions can be selected randomly, enabling the application without any further knowledge about the data. When there are additional information or boundary conditions, it can be meaningful to predefine the starting positions. The benefits of predefined starting positions are a faster convergence of the algorithm or better results. The search pattern of k-means checks for improvements by local variation, but does not consider global or random modification, which can result in local optima. The main influencing factor for insufficient solutions is the selection of the initial cluster centers. There are various approaches to tackle this issue. The geometrical optimization in the following is proceeded in a three dimensional Cartesian coordinate system with the origin in the left corner of the vehicle front at the bottom of the cars floor.

One is the k-means++ algorithm, which is an adaption of the k-means algorithm. It differs in the initialization phase as it chooses the first centroid randomly and the following ones using a distance based probability. Before placing a cluster center, the distance $D(x)$ of each data point to its nearest centroid is calculated. One of the data points is selected as a new cluster center based on a probability, which is proportional to $D(x)^2$. This procedure is repeated until all cluster centers are placed in the data set, thereby concluding the initialization phase. The further procedure is equivalent to k-means clustering.

Another option is the predefinition of cluster centers based on additional information. This requires the exact knowledge of available and meaningful positions. While developers have preferences for the position, there are limitation by diverse aspects. Three different data sets were created with the assumption, that an equal distribution of the starting points in the search space leads to good results. The distribution is mostly conducted in the x-y-plane as the spreading of the components in the x- and y-axis is significantly higher than in the z-axis. To ensure a mostly collision free search of the final location, the starting positions are located at the vehicles outer contour. The component's position will lead to a search direction towards the middle of the vehicle. To accelerate the clustering procedure, the initial z-coordinates of most centroids correspond to the mean of the z-coordinate of the entire data set. The distribution in the x-y-plane is based on an equal angle or volume separation of the vehicle.

Another approach for determining the starting points is the separation of the data set in j subsets. Each of them contains only a few data points and these are analyzed by k-means, with k* as specified cluster number, $1 < k^* \leq k$ [159]. This will generate a new data set comprising $k^* \cdot j$ data points, which is clustered once more in k clusters using k-means. The resulting position data are used as initial centroid locations for the clustering of the entire data set.

Figure 4.3 visualizes the normalized within cluster Manhattan distance, specified as $||\Delta||_1 \,/\, ||\Delta||_{1,max}$, using Euclidean and Manhattan distance as optimization criteria as well as different initial centroid positions. The three heuristic data sets are parameterized on the basis of the vehicles main dimensions and therefore are transferable to other vehicle types. After the individual cluster point initialization, the procedure is executed using squared Euclidean distance and Manhattan distance as range criteria.

The Euclidean distance of an object x is denoted as $||x||_2$, Manhattan distance as $||x||_1$. Normally k-means uses squared Euclidean distance, but as the Manhattan distance is more practical regarding cable routing, a comparison is meaningful. The results for zone numbers of one and two are not depicted due to the fact, that there is no deviation dependent on the starting positions for both. This circumstance is reasonable, as a single cluster includes all data objects leading to the same result independent of the cluster initial position. When the data objects are separated into two clusters the chance of diverse results for different starting position is low. For zone numbers from three to eight there is a difference between the range criteria. The overall best results are accomplished using the heuristic data set 3 in combination with Manhattan distance. Heuristic data set 2 performs worse than the average, indicating that not every heuristic data set

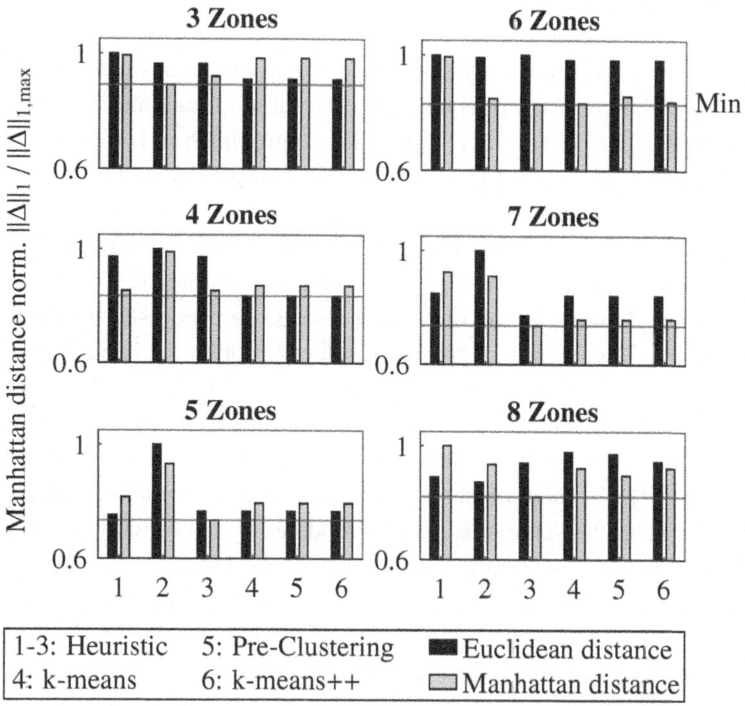

Figure 4.3: Euclidean and Manhattan within cluster distance for different initial centroid positions

is suitable for the starting positions. Limiting the scope to squared Euclidean distance as range criteria, the results of the non-heuristic approaches are better, with some exceptions. All in all the non heuristic data sets perform well using Euclidean distance for three to five zones and Manhattan distance for six to eight zones. As a conclusion it must be said that heuristic cluster initialization points in combination with Manhattan distance optimization results in the shortest overall Manhattan distance. When there is no knowledge of the optimal position and distribution of the cluster centers, one should choose k-means++ combined with the optimization of the Euclidean distance. This approach shows slightly better results compared to k-means and pre-clustering for high cluster numbers.

4.2.3 Packaging Modeling

The modeling of a physical system is dependent on the desired level of detail, therefore it is efficient to model only the relevant properties and find suitable approximations. The benefit of an appropriate approximation is the concentration on relevant aspects leading to a reduction in complexity and time needed for any calculation and modeling. This also applies to the geometric architecture analyzed within this method. Moreover the aim is to assist in early development phases, therefore it is neither necessary nor possible for the model to cover all geometric details. As the clustering results are processed by the Dijkstra algorithm in a later step, it is crucial to select a model that is compatible with both. Dijkstra's algorithm uses a graph based model, thus the vehicle is approximated by a three-dimensional grid consisting of nodes and edges. Kuchenbuch describes different methods for modeling vehicle packaging with collision avoidance [84]. As most of the package components are already pre-defined in this approach, the vehicle package is modeled by cubes with an edge length of 0.1 m, generated by volume segmentation (voxel representation). The main disadvantages of this method are the high calculation requirements caused by the large number of voxels and possible combinations. The difference in the problem scenario of this case compared to Kuchenbuch's is the significant lower quantity of components that require a placement. Most of the components, besides the traction battery cells, the ZCUs and the LV energy storage, are already pre-defined, which decreases the calculation effort significantly. A further reduction of the possible combinations is achieved by determining meaningful starting positions for these components.

Table 4.1 shows the starting position and the search algorithm to identify suitable voxels for the placement of the traction battery cells, the ZCUs and the LV energy storage. It should be noted, that the origin of the coordinate system is in the front left corner at the vehicle's floor. The placement of traction battery cells starts at the floor of the vehicle and comprises a plane limited by the vehicle's wheels. The plane is depicted in Figure 4.8 c) and is defined by the four coordinates $x_{min}, x_{max}, y_{min}$ and y_{max}. As long as the boundary condition is not fulfilled, the placement is continued with an increasing z-coordinate. The LV batteries are placed as near as possible to the mean μ_{lv} of the coordinates $\gamma(x_i, y_i, z_i)$ of all load components. Thus the search algorithm spreads spherically around the starting point while checking if the explored voxel $h(\gamma)$ is unoccupied, which is represented by $h(\gamma) = 0$. A voxel is defined as the volume element between eight vertexes. As the edges between these vertexes have the same length, the voxel is a cuboid with the edge length of 0.1 m. The search for the ZCU placement works the same way, while the starting points are the centroid positions μ_z.

Table 4.1: Starting point and search algorithm for placement of traction battery cells, LV energy storage and ZCUs in vehicle packaging

Component	Starting position	Search algorithm	Boundary condition
Traction battery cells	$x = (x_{min}, x_{max})$, $y = (y_{min}, y_{max})$, $z = 0$	$x = const.$, $y = const.$, $z \uparrow$	$\mid n_{bat} \mid < n_{bat,min}$
LV battery	$\mu_{lv} = \frac{1}{n}\sum_{j=1}^{n} \gamma_j$	$\lvert\lvert \mu_{lv} - \gamma_i \rvert\rvert_2 \uparrow$	$h(\gamma) = 0$
ZCUs	μ_z	$\lvert\lvert \mu_z - \gamma_i \rvert\rvert_2 \uparrow$	$h(\gamma) = 0$

↑ indicates the search direction. The value increases starting from the minimum value.

Figure 4.4 shows the procedure for the ZCU repositioning to ensure the compatibility with the vehicles package. The upper graphic shows the outer contour and the headlights of the input data set to better visualize the dimension of the ZCU position shift. One can identify, that the repositioning leads to minor differences of around 0.1 m, exempt zone four, which shows a difference of 0.4 m. Consequently the Manhattan as well as the squared Euclidean distance for the zones increase only little, except zone four, where the gain in both distances is significant. In such a case it can be meaningful to identify the boundary conditions leading to such an increase and try to dissolve them in early phases.

4.2.4 Reduction of ZCU Development Effort

The manufacturing cost of hardware is coupled to the quantity of products. As the number of EV manufacturers increase, cost is more than ever a decisive factor in the automotive industry, therefore OEMs need to reduce the effort needed to be competitive. As components in the vehicle differ in their com-

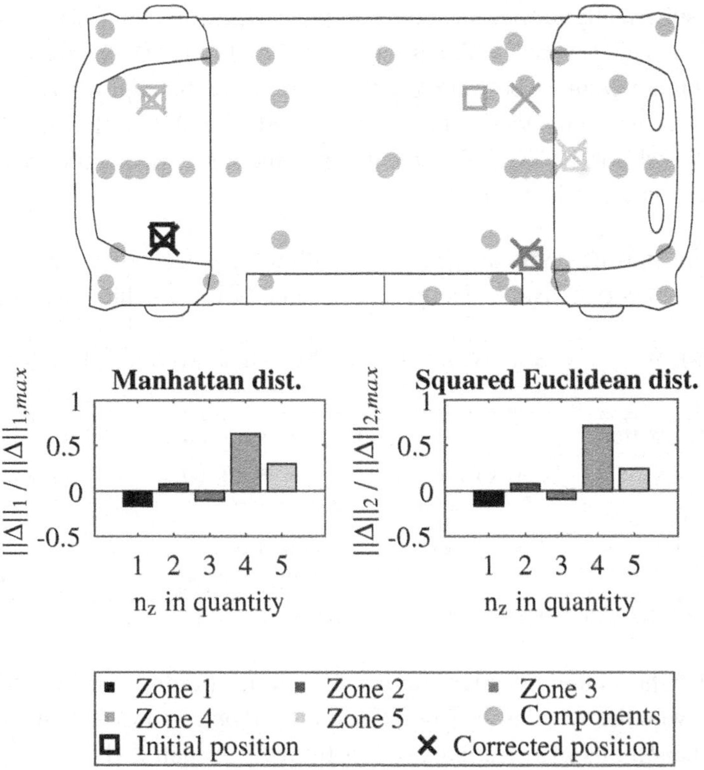

Figure 4.4: Replacement procedure for package compatible ZCU position in a five zone architecture and resulting total in between cluster distances

munication and power requirements, it is impossible to realize fully identical ZCUs, without any overhead. An approach to increase the similarity of the ZCUs is determining a identical number of connectors, resulting in a similar hard- and software architecture. As the quantity of data and power transfer might hardly differ between the ZCUs, additional modules extending the capabilities of specific ZCUs might be necessary. As k-means clustering does not realize equal cluster sizes, a subsequent resorting is done afterwards according to Maier [151]. The resorting procedure by Maier categorizes the clusters according to the relation of associated to maximum number of data points $\tilde{x}_{max}$ a cluster should comprise. This value corresponds to the maximum number of connectors per ZCU, which is defined as

$$\forall S_i \in S : \tilde{x}_{max} \leq \lceil \frac{n_{\tilde{x}}}{n_z} \rceil \quad , \qquad \text{Eq. 4.4}$$

wherein $n_{\tilde{x}}$ and n_z are the number of components and zones. Therefore, the following three groups are defined by means of the number of objects per cluster $\tilde{x}_{S,i}$:

- Receiving clusters: $\tilde{x}_{S,i} < \tilde{x}_{max}$
- Delivering clusters: $\tilde{x}_{S,i} > \tilde{x}_{max}$
- Neutral clusters: $\tilde{x}_{S,i} = \tilde{x}_{max}$

The aim of this procedure is to transform all delivering clusters to neutral ones by transferring data points from delivering to receiving cluster, thereby ensuring $\forall S_i \in S : \tilde{x}_{S,i} \leq \tilde{x}_{max}$. To fulfill this condition, an algorithm comprising three steps is developed. In the initialization phase all clusters are assigned to one of the three groups. In the second step, the distance of all elements of a delivering cluster to all centroids of the receiving group is determined. The element with the shortest distance to a cluster of the receiving group is resorted to this cluster. In a third step the clusters involved in the resorting are checked for the condition $\tilde{x}_{S,i} = \tilde{x}_{max}$. The one fulfilling the condition, is shifted to the neutral cluster group. Step two and three are repeated until there remain no clusters in the delivering group. Figure 4.5 visualizes the resorting procedure for a five zone architecture. While the initial distribution varies in the range of eight to thirteen objects per cluster respectively zone, the resorting leads to an equal distribution of eleven objects per zone.

4.2.5 Zone Assignment of Vehicle Space

As the development of new vehicle functions is more and more agile, customers will experience upgrading processes in the future to extend the initial functionality. This upgrading processes might not only require new software, but

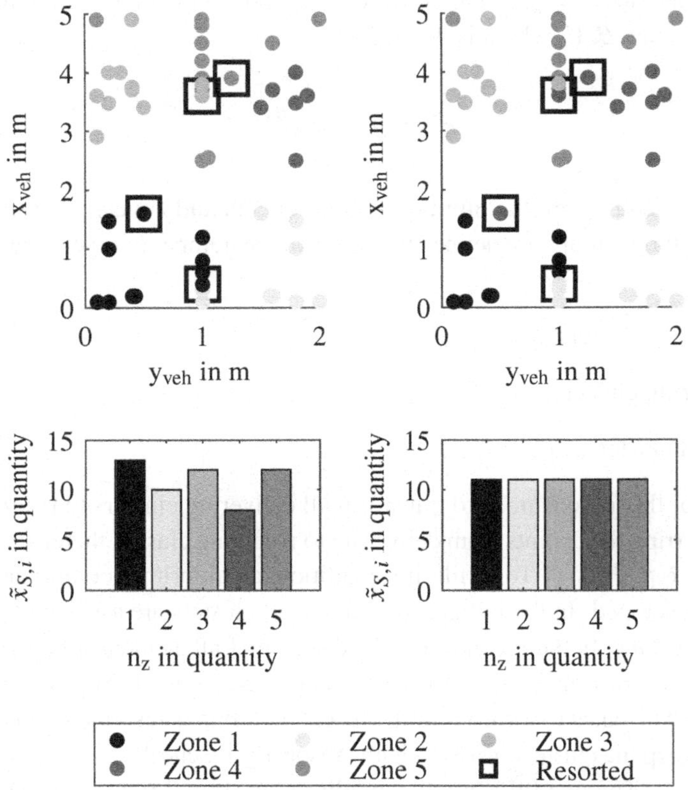

Figure 4.5: Resorting procedure for a balanced object distribution in a five zone architecture

also additional hardware in the vehicle. Therefore, it is necessary to assign the vehicle space to a zone, thereby supporting the integration process of a component. Furthermore, knowledge about the zone affiliation of the vehicle space supports modularity with respect to the reuse of entire zones in other vehicle types. There are three approaches to assign the vehicle space, respectively the voxels, to a zone, which are shown in Figure 4.6. The behavior of the approaches are visualized using the graphics on the left side and the corresponding zone alignment based on the voxel view is shown on the right side. The first is based on the components position, using them as edge points for their zone. For two dimensional environments this method is applicable without limitation, whereas in three dimensional environments it will lead to overlapping. To prevent this, zone volumes need to have the form of a polyhedron with sharp corners. These are not practicable when it comes to the reuse of zone volumes in different vehicle types and there may remain undefined voxels. Another method is the assignment of voxels to the centroids having the shortest distance to it. The drawbacks of this method are that this voxel allocation might not match with the component assignment and that the volumes of a zone have the shape of a sphere. Spheres are not as practical as a cuboid, which can be combined more easily and there is no undefined space in between, thereby supporting modularity. To tackle this problem the third method, which is a rule-based methodology, is used for assigning voxels to a zone. At first voxel sets are defined along the z-direction for each voxel in the x-y-plane. These sets are checked for the number of components that are positioned in each and the following three rules are applied:

- 1: If $\sum \tilde{x} > 1$, the set of voxels is allocated to several zones. The voxel of the components are assigned to their zones. The voxels in between are assigned to the zone of the following component voxel.
- 2: If $\sum \tilde{x} = 1$, the set of voxels is affiliated to the zone with the centroid that has the shortest distance to this component.
- 3: If $\sum \tilde{x} = 0$, the set of voxels is assigned to the zone with the centroid which has the shortest distance to the voxel in the middle of the set, representing the half of vehicle height.

The set of voxels is defined in the z-direction as literature indicates that the zone separation will mainly take place in the x-y-plane, resulting in mostly cuboid like shapes. The distance is used to assess zone affiliation, as in the cluster process for the components guaranteeing compatible results. These rules ensure that the zone volumes approximate a cuboid, having additional edges only in indispensable cases.

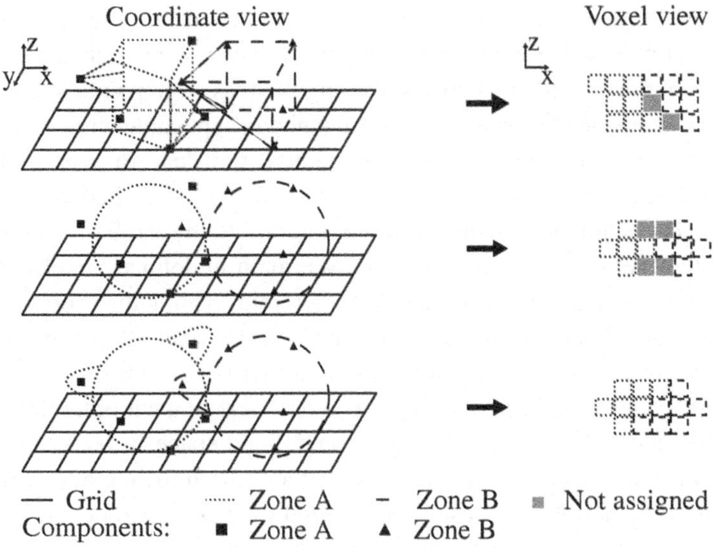

Figure 4.6: Approaches for the assignment of zones in coordinate view and their results in the voxel view

4.3 Traction Battery Optimization

The design of traction batteries for vehicles is a complex task, due to linear and non-linear dependencies of battery and vehicle. As this thesis comprises the integration of the entire EEA in the vehicle, the battery design is not discussed in detail. Approaches of other researchers who focus on optimizing the battery design in a multi-criteria environment can be used to find appropriate solutions. Therefore, an existing procedure is adapted to complement this method and model all relevant aspects of the battery. At first the NSGA-II algorithm, introduced in Chapter 3.4, is discussed in detail and applied to the problem. Consecutively, relevant aspects for the integration into a zonal architecture are discussed.

4.3.1 Traction Battery Design with the NSGA-II Algorithm

For the research in this thesis, cylindrical cells were selected as there are literature values concerning the cell number, but any cell type can be integrated in the method. At first the problem is described in more detail, followed by a short explanation of the NSGA-II algorithm. The optimization problem consists of the five functions f_1 to f_5 according to

$$\min f_1 = m_{bat} \quad , \qquad \text{Eq. 4.5}$$

$$\min f_2 = V_{bat} \quad , \qquad \text{Eq. 4.6}$$

$$\min f_3 = C_{bat} \quad , \qquad \text{Eq. 4.7}$$

$$\max f_4 = R_{veh} \quad \text{and} \qquad \text{Eq. 4.8}$$

$$\max f_5 = \begin{cases} \text{if } a_{sv} > a_{veh} & a_{veh} \\ \text{else} & 0 \quad . \end{cases} \qquad \text{Eq. 4.9}$$

Therein m_{bat} denotes the mass, V_{bat} the volume and C_{bat} the production cost of the battery cells. R_{veh} stands for the range and a_{veh} for the acceleration of the vehicle, which is compared to the specified value for acceleration a_{sv}. In order for the genetic algorithm to produce meaningful results, the dimension can be reduced using the aggregation tree method [160]. Since f_1, f_2 and f_3 aim to minimize the number of battery cells, they are combined into $\mathcal{F}_1$. f_4 and f_5, on the other hand conversely maximize the number of cells, thus they are combined into $\mathcal{F}_2$.

$$\min \mathcal{F}_1 = \lambda_1 \cdot f_1 + \lambda_2 \cdot f_2 + \lambda_3 \cdot f_3 \quad , \qquad \text{Eq. 4.10}$$

$$\max \mathcal{F}_2 = \lambda_4 \cdot f_4 + \lambda_5 \cdot f_5 \quad , \qquad \text{Eq. 4.11}$$

where λ_1 to λ_5 is the weighting of the factors in the objective functions. The aggregation of the functions is supported by the aggregation tree, which is provided in Appendix B.1.

The solution space is limited to reduce the computation time. The upper boundary is formed by typical values from series vehicles, such as the 7104 cells in a Tesla Model S 100 kWh and a lower boundary derived from the demanded range [133]. In addition the computation time is limited by a termination criterion. By using genetic algorithms, the resource usage is reduced and the solution quality is improved. The Non-dominated Sorting Genetic Algorithm II (NSGA-II) is a genetic algorithm for solving multi-objective optimization problems and shows the best results with two to three objective functions [161, 162]. The algorithm designed by Deb et. al involves six steps [163]. First, an initial population is randomly generated, which is sorted into fronts. A front comprises a set of individuals dominated only by elements of better fronts. The best front consists of the Pareto optimal solutions. Pareto optimality is accomplished when it is not possible to improve any characteristic of a solution without decreasing another. Within each front the crowding distance is calculated, which is the Euclidean distance of all properties of an individual to the neighbors within the same front. Individuals are selected based on the fronts and crowding distances, thus better fronts and larger crowding distances are accepted, while the rest are discarded. In the next step, further individuals are generated by means of crossing and mutation. Finally, the population is reduced to the initial size through a fitness value based selection. The result is a Pareto front representing all non-dominated solutions. Domination in this context means, that a solution is better in one criteria, but at least equally good in all other criteria.

To select the best solution from the Pareto front, there are several approaches. One universally applicable method is to use the Euclidean distance to the optimum [164]. Since the normalized cost should be minimized, the zero point of both axes is the optimum. Another method is the specific selection by an expert. This expert weighs the two criteria based on his experience resulting in the minimum of the weighted sum of both criteria. The weighting is selected with respect to the vehicle type. For a public shuttle, cost is the decisive factor, whereas for a sports car it is performance.

4.3.2 Parameterization of Genetic Algorithms

The genetic algorithm searches the optimal number of battery cells for a specific vehicle configuration and its electricity consumption. The parameters used for the optimization are essential to ensure an efficient and reliable search process. Crucial parameters are population size, crossover and mutation function as well as the termination criterion. For the population size a number of 50 is selected as this forms a compromise between the needed calculation time, the number of variables and the quality of the solution. The crossover function uses a crossover rate of 0.8 and combines two objects from the parent generation with a random factor. As the quality of the generation changes during the optimization procedure, the mutation function must adjust to it. Therefore, an adaptive mutation function is used which changes the step size depending on the change in the spread of generations and the mutation rate decreases linearly with the progress in the calculation process. The optimization procedure stops after a predefined number of generations is reached or when there is no more improvement for a specific number of steps. Although the parameters are crucial for the quality of the solution, as well as for the computation time, it is not possible to identify the optimal parameter set for all inputs. The aim is to reliably find good results within a reasonable time for all types of input parameters. For the case that the range specification is a hard requirement, the minimum number of necessary cells to accomplish this range is set as a lower boundary to the solution space. Further details regarding the settings are listed in Appendix C.4.

4.3.3 Battery Cell Placement Strategy

The position of traction batteries has a decisive influence on the entire vehicle packaging as well as on the center of mass. There are two commonly used battery shapes in current EVs, which are the 'T-shaped'- and the 'floor'- configuration [165, 166]. The T-Shaped configuration is located along the longitudinal axis and in the transverse direction in front of the rear axis. Therefore, from a top-down view on the vehicle platform it corresponds to a T. This brings benefits for safety as the battery is not in the crash zone, but it needs to be higher as the area is proportionally small, resulting in less space for the cabin. The floor configuration uses almost the full area between the wheel axles, resulting in a lower battery height and therefore in more cabin space. As the side areas are used, the battery is part of the crash zone, thereby increasing the chance of a battery damage in case of a crash. The aim of autonomous electric shuttle buses is to transport as many persons as possible and must offer standing room for some use cases. To meet this goal, cabin space is essential and needs to be maximized if possible, therefore the floor configuration is more useful. The safety risks for the floor variant are lower for the use case of autonomous shuttle buses, since they move at lower speeds and mainly within cities.

The position algorithm is developed with consideration of existing concepts for future vehicles, which are shown in Figure 4.7. The upper graphics show the chassis and the lower ones the corresponding prototypes of four vehicle concepts. They are designed for the transport of people for different use cases. The vehicle in Figure 4.7 a) and Figure 4.7 c) offer seats and hence are suitable for commuting, while the ones in Figure 4.7 b) and Figure 4.7 d) feature standing room, enabling a meaningful operation for public transportation over short distances. In all chassis the batteries are positioned in the area of the vehicle floor between the two axles. The room around the axles in the radius of the wheels is reserved for the same thing and the space at the vehicle's front and end is dedicated for further electronic components such as AD and ADAS sensors, auxiliaries and PT components. Under these boundary conditions the placement algorithm identifies the plane that is available for placing battery cells, which is defined by the vehicle width y_{min}, y_{max} and the distance between axles and the radius of the wheels $x_{min} = x_{axle} + r_{whl}$ and $x_{max} = x_{axle} - r_{whl}$. The batteries are placed in several layers starting from the bottom, while a layer is defined by the height of the battery cells. The cells are placed next to each other with no offset. Although an offset increases the packaging efficiency when using cylindrical cells, the abstinence of an offset enables the usage of all cell types. The total number of cells is divided by the number of cells per plane to identify the completely filled layers. The remaining cells are symmetrically placed around the vehicle center to realize a symmetrical weight distribution for vehicles featuring four wheel hub drives. The weight distribution can be varied for vehicles having one driven axle by weight transfer in the direction of the driven axle.

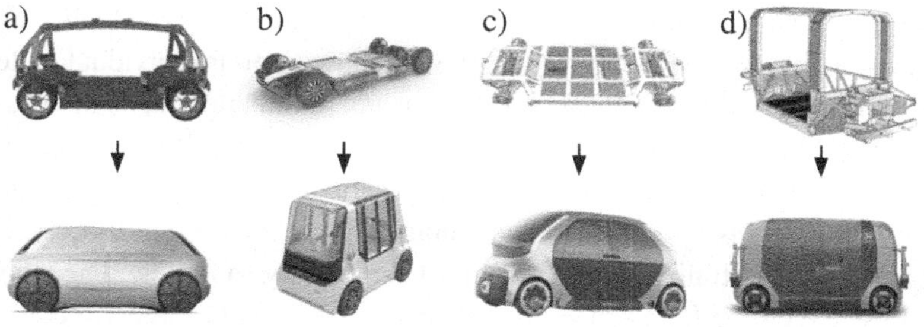

Figure 4.7: Vehicle chassis and concepts: a) Urban Modular Vehicle [167], b) Schaeffler Rolling Chassis and Schaeffler Mover [168], c) FlexCAR [169] and d) UNICARagil autoSHUTTLE [170]

4.3.4 Battery Module Separation Strategy

The fully autonomous operation of the individual zones would be beneficial especially for the safety behavior of the EEAs in critical situations. While the isolation of the data network places additional requirements on the functional architecture, the power supply does this for the hardware. The power supply concept of the UNICARagil platform, introduced by Goth, covers four batteries, each in a separate housing and assigned to one zone [34]. It is able to disconnect a zone and its assigned components electrically from the remaining powernet, allowing autonomous operation. While this concept features great benefits regarding functional safety, an increase in the battery number causes disadvantages in assembly, production and system complexity. To avoid these disadvantages the battery designed by this algorithm combines all cells in one housing, but assumes the assignment of modules to the zones in a predefined static topology. Lee et al. as well as Tippe et al. have proven the fundamental feasibility of this approach using a prototype setup comprising a few battery cells and via simulation [171, 172]. A dynamic reconfiguration is under research, but not realized in series vehicles today and therefore not assumed [173].

Figure 4.8 a) shows the circuit for a ring topology of a zonal power supply, enabling the isolation of single zones. By opening the switch of an electric path, the zone is separated from the rest of the power supply system, thereby isolating errors in this specific zone. This topology is implemented in the UNICARagil vehicles, which is indicated in Figure 4.8 b), showing four individual batteries. The algorithm used in this thesis for the placement of battery cells unites all of them in one housing, which is visualized in Figure 4.8 c). The assignment to zones is realized by the allocation of battery modules and the usage of multiple battery plugs. This approach has advantages for packaging and safety, but it imposes two constraints on the design of the battery modules. The smallest battery size required for a zone must not be smaller than one battery module and the capacity of every zone must be equal or an integer multiple of the smallest module's capacity.

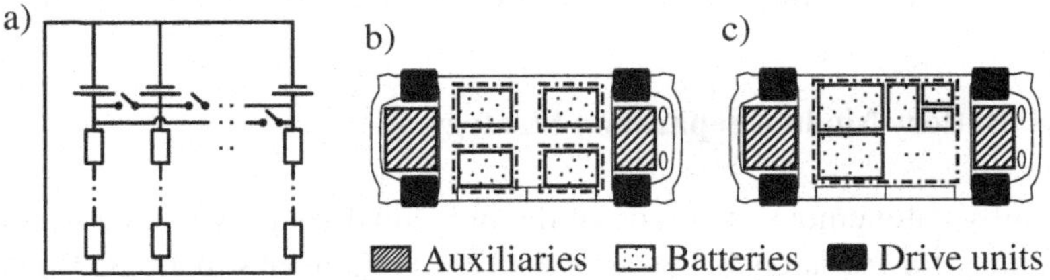

Figure 4.8: Battery separation for zonal power supply, a) circuit of a battery ring topology, b) UNICARagil battery placement and c) algorithm battery placement

4.4 Low-Voltage Electric System Design

As defined in Figure 4.2, several zones share one LV battery and one SC. The more batteries and SCs are used, the greater the reduction of the total wiring harness length and the total weight and volume of the energy storage increases. The LV energy storage is relevant to all safety-critical electric components as it is indispensable for the redundant power supply for these components. Therefore, another k-means clustering is needed to identify the best number and position of the LV energy storage. Table 4.2 shows the approximate cost for the usage of one to eight LV batteries and SCs for the autoSHUTTLE and the given data set. The cost comprise the energy storage and the dedicated cable to the components and are determined using cost functions introduced in Chapter 3.5. The length of the cables correspond to the Manhattan distance between the cluster centers and the components position, while cable cost are determined using the table in Appendix B.1. The cost for the LV batteries rise with each additional unit, while those of SCs show a minimum for two and from that point on continuously increase. The minimum is dependent on the data set, but greater dispersion will shift the minimum to higher energy storage numbers. Additional to the cost, there are functional aspects that need to be considered. A number of two or more LV batteries or SCs improves functional safety, as functions along with their dependencies can be separated on different power sources. Consequently, the number of these are specified as two. The assignment of the components to the LV energy storage is based on the clusters, while their centroids correspond to the storage. The position for them follows the same procedure as used for the search of the ZCU position, however there are two additional constraints. The first is, that the needed storage for both can strongly vary in size, therefore a search is made for a space that offers enough space in all three dimensions. The second is that their weight can be much higher than those of the ZCUs, thus positions near to the vehicles center are prioritized.

Table 4.2: Total cost in EUR for cables and energy storage depending on the number of energy storage units

	1	2	3	4	5	6	7	8
LVB	57.23	100.41	146.67	194.18	239.47	288.01	333.72	381.06
SC	17.59	13.91	14.35	17.12	18.25	20.78	23.60	26.44

4.5 Vehicle Wiring Harness Optimization

Important aspects of the wiring optimization are considered in the following sections. Therefore, the requirements transfer from vehicle or system level to the path planing, presented in Section 3.3, is explained as well as the procedure to adjust the cable bundling. Based on the optimized routing core properties of individual cables and structural properties of wiring harness are determined.

4.5.1 Requirement based Path Cost Modification

Apart from the length and weight, there are further criteria that must be considered in wiring harnesses development. For example, high temperatures affect the lifetime of cables and Electromagnetic compatibility (EMC) may cause functional limitation. The consideration of requirements in the path finding step is done by separate edge cost components. To realize an uniform edge cost factor, the sum of edge cost components is normalized. The cost for an edge n result from

$$W_n = \frac{1}{k} \sum_{i=1}^{k} w_{n,i} \quad , \qquad \text{Eq. 4.12}$$

with $w_{n,i}$ as the cost component for edges n, where k is the number of requirements. Each cost component ranges from zero to one. Cost components are determined using functions that assign appropriate cost for each path. One of these functions evaluates each edge according to its distance from the main cable routes. These are cable trays that reduce manufacturing and assembly cost by bundling individual cables. According to Neckenich, there are three typical designs of main cable routes, these are the E, H and double-H structure [174]. They comprise the routes in the x-y plane, and the z-axis is neglected. While this was suitable in the past, new vehicle concepts with higher dimensions in the z-axis may require an extension by this dimension. Additionally, there is an increasing number of loads in higher positions, such as cameras and lidar sensors, as well as new use cases like autonomous shuttles, which have designs extended in the z-axis [175]. Figure 4.9 visualizes the concepts for main cable routes according to Neckenich as well as an extension in the z-axis.

For edge i, the cost component of the requirement 'follow the main cable routes' results from the distance to them according to

$$w_{n,i} = \begin{cases} 2\,w_1 & \text{for } w_1 < 0.5 \\ 1 & \text{for } w_1 \geq 0.5 \end{cases}, \qquad \text{Eq. 4.13}$$

using

$$w_1 = \min \sqrt{(x_i - v_x)^2 + (y_i - v_y)^2 + (z_i - v_z)^2}\ , \qquad \text{Eq. 4.14}$$

wherein x_i, y_i, z_i are the inspected coordinates and v_x, v_y, v_z are the coordinates of the main cable routes.

The edge cost W_n for an autonomous shuttle vehicle are shown in Figure 4.10. A double-H structure is specified for the main cable routes. This is a common approach for long and large vehicles, thus it is most suitable for the auto-SHUTTLE with a length of almost five meters. The main cable routes can be identified by the lower edge cost in the floor and roof of the vehicle.

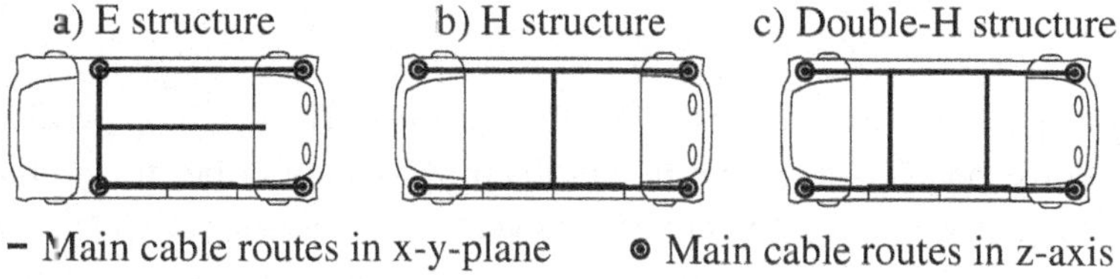

Figure 4.9: E, H, and double-H structure for the main wiring cable routes, adapted from [174]

4.5.2 Cable Bundling

Cable bundling can broadly be defined as the concentration of separate cables into a single strand. Bundling comes with additional requirements which tends to increase the overall wiring harness length. In contrast it significantly reduces the effort needed for production and installation, simplifies integration and reduces complexity of the wiring harness as a whole. Some of these aspects can be associated with cost, but most can not, hence it is hard to

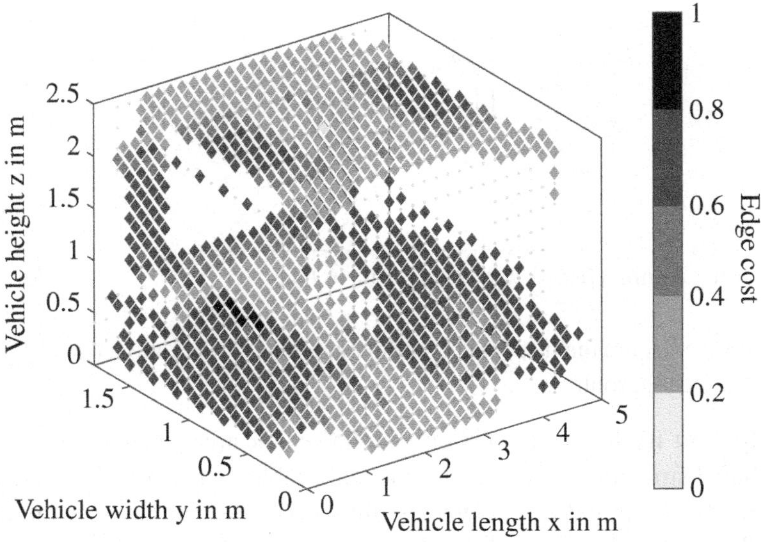

Figure 4.10: Edge cost of an automated shuttle bus for shortest path problems, adapted from [151]

determine the optimal rate for the bundling process. The bundling problem in a graph corresponds to a Steiner tree problem, which describes the problem of identifying an optimal path between a given set of objects, which are referred to as terminals [176]. The optimal path connecting them is characterized by the lowest total weight of the connecting edges. As most connections are in-between components of a zone, the problem can be split into separate zone wiring harness optimization problems. Therefore, the number of variants is reduced, resulting in shorter computation time.

Siegle defined an iterative algorithm to optimize the bundling in zonal wiring harnesses, which is visualized in Figure 4.11 and described in the following section [177]. The initial wiring harness is designed by connecting all components in one strand as visualized by Figure 4.11 a). The strand starts from the ZCUs and is connected to the closest components, in this case it is A. In the next step this component is connected to another one of this zone, which was not previously connected to the strand, here component B. This procedure is continued until all components of a zone are connected by one strand. The procedure of Siegle is in the following described by two steps. Starting with the last component of the strand, its connection is shifted along the entire strand until it reaches the ZCU and the variant having the lowest edge cost is selected as indicated by Figure 4.11 b) and Figure 4.11 c). The pool of variants comprises the initial configuration as well as the direct connection from the ZCU. This procedure is repeated for every component for which the order is determined by their position along the strand, starting from the end of the strand. It should be noted that the newly created paths are also checked for shorter path variants. The second iteration step is utilized whenever there are new breakout points created, which applies when the edge cost for a bundled path is lower than for an isolated path as illustrated in Figure 4.11 c). The created breakout point is shifted in the direction that reduces the edge cost the most. The shifting process is continued until there is no cost reduction achieved anymore. Figure 4.11 d) shows that the breakout point is moved to the left until a position is reached where no further improvement can be realized. An alternating use of iteration step one and two leads to the best results as they optimize configuration and geometry uniformly, which results in the configuration visualized in Figure 4.11 e). Furthermore, Siegle defined two procedures for the optimization of zone wiring harnesses. One of them optimizes the HV level first and subsequently the LV level, the other optimizes both in one process. In a parameter study he identified that the first approach results in a wiring harness that has lower cost and is shorter [177]. The second procedure leads to lower length expansion of the zone wiring harness. Both variants are meaningful, but the first is used in all simulations in this thesis as the automotive sector is highly cost driven.

The most important parameter for the algorithm when evaluating paths is their cost. It is modified by the bundling factor b_f, a parameter determining the additional bundling cost when a cable is bundled with others on the same path. It is defined as a function of the cable number per edge b_n according to

$$b_f = f(b_n) \quad , \qquad \text{Eq. 4.15}$$

while $f : H \rightarrow \mathbb{R}$ applies to the function f wherein H is its domain. The root function is a meaningful example as the growth of the edge cost per cables decreases, which corresponds to practical application. The influence of the bundling factor can be exemplified by the lower and upper limits. A

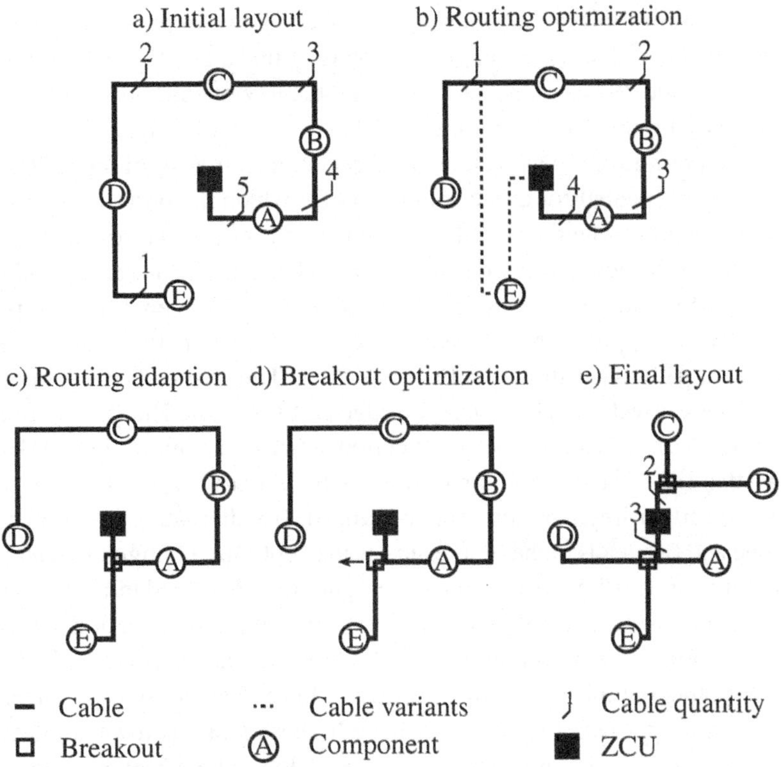

Figure 4.11: Iterative cable bundling in rectangular weighted graphs, adapted from [177]

bundling factor of infinity will result in minimal bundling, leading to a wiring harness consisting of the maximum number of individual cables, each following its shortest path. In contrast, a bundling factor of zero causes a maximum bundling in combination with high individual length increases. Consequently, the selection of an appropriate bundling factor is essential.

To highlight the influence of the bundling factor, the effect on crucial properties of the wiring harness are listed in Table 4.3 for $b_f = \sqrt{b_n}$ and $b_f = 0.7b_n$. The values are determined by a simulation of a five zone architecture in the autoSHUTTLE. The usage of linear functions reduces the bundling length

$l_{bnd,wh}$, the total length l_{wh} and the mass m_{wh} of the wiring harness, while it increases the expansion $l_{e,wh}$ as well as the number of angles A_{wh} and breakouts B_{wh}. Thus the root function leads to a more intense bundling, whereas the linear functions tend to use shorter and more direct paths. The difference in each property is listed as Δ_{bnd}, indicating an influence of between 3.78 % to 23.1 %, emphasizing the impact of the bundling factor.

Table 4.3: Influence of the bundling factor on wiring harness characteristic properties for a five zone architecture of the autoSHUTTLE

	$\sqrt{b_n}$	$0.7b_n$	Δ_{bnd} in %
$l_{e,wh}$ in m	38.40	43.80	+14.0
$l_{bnd,wh}$ in m	193.2	184.0	-4.80
l_{wh} in m	352.9	335.3	-4.99
m_{wh} in kg	8.458	8.139	-3.78
A_{wh} in quantity	186	229	+23.1
B_{wh} in quantity	67	79	+17.9

$\Delta_{bnd} = \frac{0.7b_n - \sqrt{b_n}}{0.7b_n}$

Figure 4.12 shows a patent by Tesla Inc., which maximizes the bundling by using a backbone network for data and power, which comprises two cable trays along the longitudinal axis and thus is a practical approximation of zero for the bundling factor.

In this chapter it has been demonstrated that there is no optimal bundling factor for all applications, on the contrary it is application specific, thus it must be identified for each case with respect to the manufacturing and production parameters. While Table 4.3 already foreshadowed some properties of wiring harness, the following chapter contains more details and further properties are introduced.

4.5.3 Determination of Wiring Harness Properties

The data traffic in automotive networks has undergone an abrupt increase after assisted driving and connectivity functions were introduced in the 2010s.

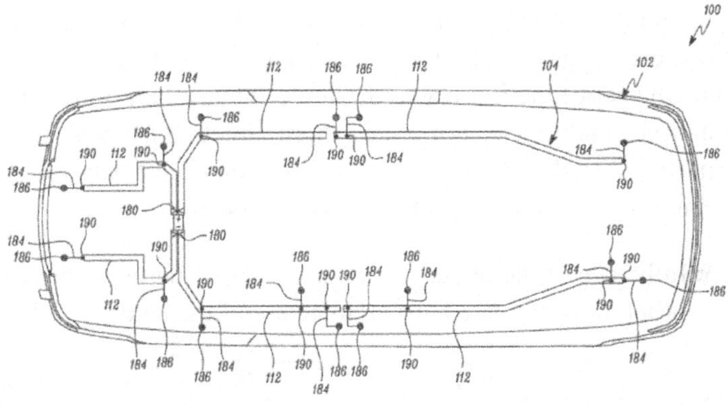

Figure 4.12: Patent for a *Wiring System Architecture* by Tesla Inc. [118]

Further expansion of these and other functions keep this trend ongoing, consequently new data transmission technologies such as 'Automotive Ethernet' were established. On the one hand it is crucial to ensure compatibility with the increasing amount of traffic, therefore all data cables in this thesis use 'Automotive Ethernet'. On the other hand it is meaningful to adjust the data rate based on expected traffic to determine the material cost accurately, hence the data rate of the backbone is limited to 1 GB/s (1000BASE-T1), while the data rate between the network nodes and the ZCUs is defined as 10 MB/s (10BASE-T1). Based on the technology cost factor d_c and mass factor d_m ,which are both defined per meter, it is possible to determine both properties for each cable separately. The energy transmission is carried out using copper cables and their diameter is specified in accordance with the maximum current given in Appendix B.1, which is based on DIN EN 60228 respectively VDE 0295:2005-09. Cost e_c and mass e_m factors for automotive typical electric cables are listed in the same table, allowing an accurate calculation to be made for both. With rising voltage, the effort needed for insulation and mounting typically rises, therefore a cost factor for the voltage level e_{vl} is added to consider these. There are mainly two variants for the design of power cable paths in machines. The first is to design each wire individually, insulate them, and bundle them into another insulator material, leading to a reliable but manufacturing-intensive wiring harness. The second variant is only meaningful if the wiring system has many bundled paths, as it unifies them in a common conductor. This variant comes with lower manufacturing exertion, as the number of cables is reduced drastically. The upgradeability of the systems is also improved, as the system can include a plug-and-play connector for additional components. A major disadvantage of the second variant are the disproportional heavier and thicker cables, as the current per cross section rate decreases for higher diameters. Another disadvantage related to the upgradeability is the necessity to design the common path with additional buffer to enable hardware upgrades, resulting in further extra volume and weight for the wiring harness. Although the second variant has significant potential, a meaningful application in vehicles is related to the platform and their corresponding strategy. Consequently, the established and generally applicable first variant is chosen for the cable design of the energy transmission.

Properties such as cost, length and mass characterize a wiring harness fundamentally. Another significant aspect is the structure of the wiring harnesses itself, comprising properties such as angles, breakouts and length expansion. Any shift in direction equal or higher than 45° is considered as an angle causing additional cost for vehicles and wiring harnesses. Furthermore, it is essential to comply with the bending radius as an undercut can result in damaged cables. Although the bending radius is not considered in the calculation of the physical properties, the compatibility with standards is ensured and described in detail in Appendix D.1. A breakout occurs whenever a cable enters or leaves a common path with at least one other cable. The number of breakouts is a measure for the quality of bundling, but disregards the bundling lengths. In combination with the bundling length it provides a ratio that indicates the efficiency of a bundling process. Length expansion is the length of all wiring harness paths, while common paths are only counted simple. It is an indicator for the expansion of wiring harnesses and thus also for the limitation for the packaging and other systems. The physical properties complemented by these three characteristics enable an evaluation of not only the wiring harness itself, but also the corresponding efforts for the packaging and the entire vehicle architecture.

Although the optimization of the wiring harness paths towards the shortest routing brings benefits, especially with respect to the physical characteristics, it results in the neglect of other aspects. In the following, some of these aspects are named without claiming completeness. These include available space and accessibility for mounting the wiring harness at specific positions, adaptability for vehicle variants, production cost, especially regarding the ratio of required automated and manual work. Furthermore, it is not considered whether a change in the vehicle frame, respectively geometry, is more cost-efficient than changes in the wiring harness. Although the adaptation of the path cost enables the consideration of various requirements, the transformation poses additional challenges and will not be possible for all aspects. Consequently, the solution accomplished by this method can be optimal in terms of wiring harness length, but other stakeholders may demand adaptations or changes to make the solution realizable from their perspective.

5 Case Study and Discussion

The action previously planned and taken resulted in a method, which is evaluated in this chapter. The aim is to generate an understanding of the suitability of the method, therefore a case study is conducted and the different solution variants are discussed. The case study consists of two vehicle data sets and four modifications of these, which ensure the generality of the method. At first two vehicles and their data sets are introduced and the variants are evaluated using the proposed method. In the subsequent sections the effect of characteristic modifications on this method are verified. Verification is the review of a product, or in this case the scientific method, whether it fulfill the initial defined targets and therefore its development was done correct. It should not be mixed up with validation, which is the check whether a product fulfills the users needs and thus the correct product was developed. The validation is conducted in chapter 6 by evaluating the overall method with the initial research questions.

5.1 Simulated Vehicles

This sections provides a brief overview of the simulated vehicles and their characteristic properties. The aim is to utilize data of existing vehicles to ensure the practicality and usefulness of the proposed method. The necessary data comprises extensive information about the electronic components and the geometry of the vehicle. While the outer shape is well-known for all vehicles, the detailed vehicle geometry as well as spatial and electric component data are not available making it hard to compile a complete data set. The first set is based on one of the vehicles of the UNICARagil platform. While, in particular, the inner geometry is almost similar to the real vehicle, adaptions are made regarding the components power to approximate realistic values in series-production vehicles. The second set uses the technical data of a premium segment Sport utility vehicle (SUV), wherein the components position is fully and the size partly known. The electrical data are approximated using data

J. Maier, *Optimization of Zonal Electric/Electronic Architectures Using a Coupled Modeling Approach of Package and Wiring Harness*,
Wissenschaftliche Reihe Fahrzeugtechnik Universität Stuttgart,
https://doi.org/10.1007/978-3-658-48188-9_5

sheets or literature values. Both data sets are presented in Table 5.1. The use of the second data set serves the purpose of validating the methodology and not to indicate how the vehicle should have been built. This would not be meaningful either, as many parameters are not known and not all aspects of real vehicle development processes can be considered within the scope of this work. For this reason, the exact vehicle type is not specified. The number of electronic components is almost similar to the data sets of real vehicle used by Mauser and is therefore able to represent recent vehicles [178].

Table 5.1: Vehicle characteristic properties

	autoSHUTTLE	Premium SUV
Vehicle class	Shuttle	SUV
Technological maturity	Prototype	Maturity phase
Passenger number in quantity	8	5
Desired range in km	100	331
Kerb weight in kg	2500	2560
Height in m	1.9	1.633
Length in m	4.97	4.915
Width in m	1.85	1.937
Drive units in quantity	4	2
Drive power in kW	57.6	183
Electronic components in quantity	55	50
Average idle consumption in kW	8.195	10.555

5.1.1 autoSHUTTLE

The autoSHUTTLE is a prototype shuttle bus developed in the UNICARagil project. The aim of the project was to develop a modular platform comprising the domains geometry, mechatronics and software in order to establish an architecture for future automated EVs. The autoSHUTTLE is one of the vehicles developed on the basis of this platform with the purpose of tackling the high volume of traffic in cities as well as the insufficient availability of public transport in rural areas. Therefore, it resembles other autonomous shuttle buses in appearance and differs from typical passenger vehicle by its outer shape and proportions [179]. Its most characteristic properties are the seats and standing room for up to eight persons, a comprehensive Human machine interface (HMI) comprising two displays and an app as well as the other technical features of the UNICARagil platform, such as automated driving functions of level 4 and wheel hub drives. Figure 5.1 shows the side and the top view of the autoSHUTTLE.

a) b)

Figure 5.1: autoSHUTTLE: a) Side view, b) top view, adapted from [170]

Based on the use cases of automated vehicles, the platform was supposed to comprise four wheel hub drives, enabling lateral driving maneuvers and four sensor hubs for environment perception. The function of drive units and optical sensors restrict the selection of their position, resulting in a data and electric power centroid in each corner of the vehicle. Consequently, the project team defined four zones for the platform by dividing it along its longitudinal and transverse axis. A weakness of this assumption is the arrangement of other electric components as well as the wiring harness complexity and cost in the decision process.

5.1.2 Premium SUV

The premium SUV data set corresponds to a state-of-the-art vehicle of the class SUV. SUVs are one of the most favored vehicle classes in China, Europe and the United States, and their share in the total number of cars worldwide is increasing [180, 181]. Typical for the vehicle class SUV is their large height and length, increasing the wiring harness characteristics such as length and mass. Furthermore, a premium SUV comes with many comfort and safety functions resulting in a high number of electronic components. The complexity of EEAs and wiring harnesses gains by an increasing number of electronic components, consequently a complex and well equipped vehicle is more appropriate to show the advantages and disadvantages of different zone variants. The used SUV is driven by two electric axle drives, has assisted driving functions of level 3 and comprehensive comfort and safety functions. Therefore, its technology is state-of-the-art and the EEA design is of higher or equal complexity to the majority of most vehicles.

5.2 Zone Number Variation

The suitability of an EEA in this thesis is determined using characteristic properties which are derived from literature and technology trends. The wiring harness is an essential component in the design process of zonal EEAs and the aim is to reduce its complexity and length. Therefore, both the characteristic attributes of EEAs and wiring harnesses are discussed in the following section. First they are analyzed individually for each vehicle, second the results of both vehicles are compared and evaluated. It should be considered, that the values for the one zone architecture are for reference and should not be seen as part of the solution space. Nevertheless, they are presented whenever a determination was possible.

5.2.1 autoSHUTTLE

Figure 5.2 shows the characteristic properties of the autoSHUTTLE's EEA determined by the method. On the left side are the core attributes including mass, material cost and wiring harness length, on the right side are function based properties comprising electrical balance, flexibility and SOF. The mass of all architectures is in the range of around 430 kg to 440 kg. In particular, the increase from one to two zones reduces the weight drastically, while a further increase of the zone number up to eight results in slight variations in a range of 2 kg to 3 kg, with a local minimum for three zones and a global minimum for eight zones. As the traction battery has a mass of at least 300 kg, improvements in the wiring harness result only in minor weight reduction with respect to the total mass of the EEA. The material cost are also dominated by the vehicle battery's cost of at least 5000 EUR. The minimum is accomplished for three zones, due to a relatively low wiring harness weight and low ZCU cost. Although the wiring harness has lower cost for higher zone numbers, the increasing ZCU cost counterbalance them resulting in overall rising cost starting from three zones. The wiring harness length undergoes a heavy drop when a second zone is introduced. Higher zone numbers lead to a further decrease of the wiring harness length, while a zone number of three again indicates a local optimum. Additionally to the physical values, there are further properties that characterize functional and architectural aspects. The electric balance indicates how good these zones are balanced with respect to electrical energy and power. One zone embodies the optimum, as an architecture of one zone has no other zones for comparison. While most zones show values around 0.75, the four zone architecture lead to a slight better result, justified by the four wheel-hub drives in each vehicle corner. Flexibility indicates the sensitivity of architecture in terms of hardware changes and decreases with rising zone number, while seven zones are associated with a minimum. The SOF demonstrates the availability of functions for the case of a breakdown of the most powerful zone. While it is self-evident that the value of this property increases with higher zone numbers as the division becomes finer, it is remarkable that the highest value is already reached with five zones and remains at this level with one exception.

The simulation results for the wiring harness of the autoSHUTTLE are depicted in Figure 5.3. The characteristic properties of the wiring harness comprise three core qualities on the left side, two functional attributes on the right side as well as another structural characteristic. Although the wiring harness length was already discussed in the context of the entire EEA, it is meaningful to discuss its influence on the wiring harness. Mass and material cost show a sharp drop for zone numbers of two to eight due to the length reduction of the wiring harness. While the optimum for all three attributes is accomplished for eight zones, the three zone architecture again indicates a local minimum in the range of two to four zones. On the one hand it seems conclusive that the high zone number reduces the cable length, resulting in a mass and cost reduction. On the other hand it is remarkable that a three zone architecture shows advantages compared to higher numbers due to an optimal clustering.

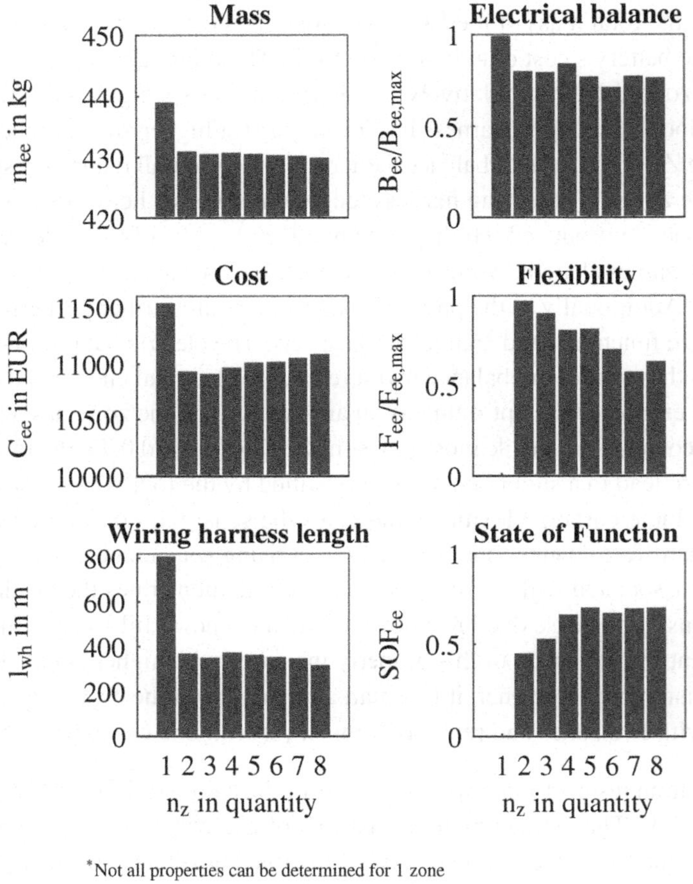

Figure 5.2: autoSHUTTLE EEA characteristic properties for zone numbers of one to eight

The complexity balance demonstrates the difference in the complexity of zone individual wiring harnesses showing a maximum at four zones. In the range of two to five it is the case that the stronger the number differs from four, the worse is the balance. This is linked to the component distribution in the vehicle. It was designed to have four zones, respectively equally balanced data and zone hubs; therefore, the component distribution is relatively similar in four zones resulting in analog wiring harnesses. In contrast the summed complexity of the entire wiring harness shows a continuously rising complexity due to an increasing number of components and interfaces. The complexity of the individual wiring harnesses can be determined by the gain of the total complexity. The relatively small gain indicates a decreasing complexity of the zone wiring harnesses. The length expansion represents the total length of all edges occupied by the wiring harness in the vehicle. While an expansion of the length with increasing zone number is reasonable, the local minimum for six zones indicates an efficient bundling, as the bundling coefficient is constant for all simulation results shown in Figure 5.3.

5.2.2 Premium SUV

The premium-segment SUV's characteristic properties are depicted in Figure 5.4 and Figure 5.5. The vehicle mass decreases as the number of zones rises, but there are pairs of almost similar results, such as two and three zones. These pairs are recognizable in the trend of the material cost albeit with slightly different relations with respect to a zone number of two and six. The wiring harness lengths correspond to the ideally expected trend level and relation. The length values decrease continuously until a certain number of zones is reached, for this vehicle the number is seven. From that point on, any increase of the zone number will bring no further benefits with respect to the wiring harness mass or length. The electrical balance decreases continuously with two exceptions. Zone numbers of six and eight show relatively good results, probably as a result of the different starting point determination for the clustering. Flexibility for zonal approaches applied to this vehicle leads to good values for zone numbers of two and three, while more zones result in a flexibility value of around 0.7. The graph shows a steady increase in the SOF of the architectures up top five, followed by a drop, which is constant for the remaining variants. The pairings in mass and material cost of the EEA are caused by the wiring harness characteristic as these pairs can be identified in the corresponding graphs in Figure 5.5.

The wiring harness length does not show the same pairing as the EEA, therefore the additional costs are caused by further cost components such as breakouts and

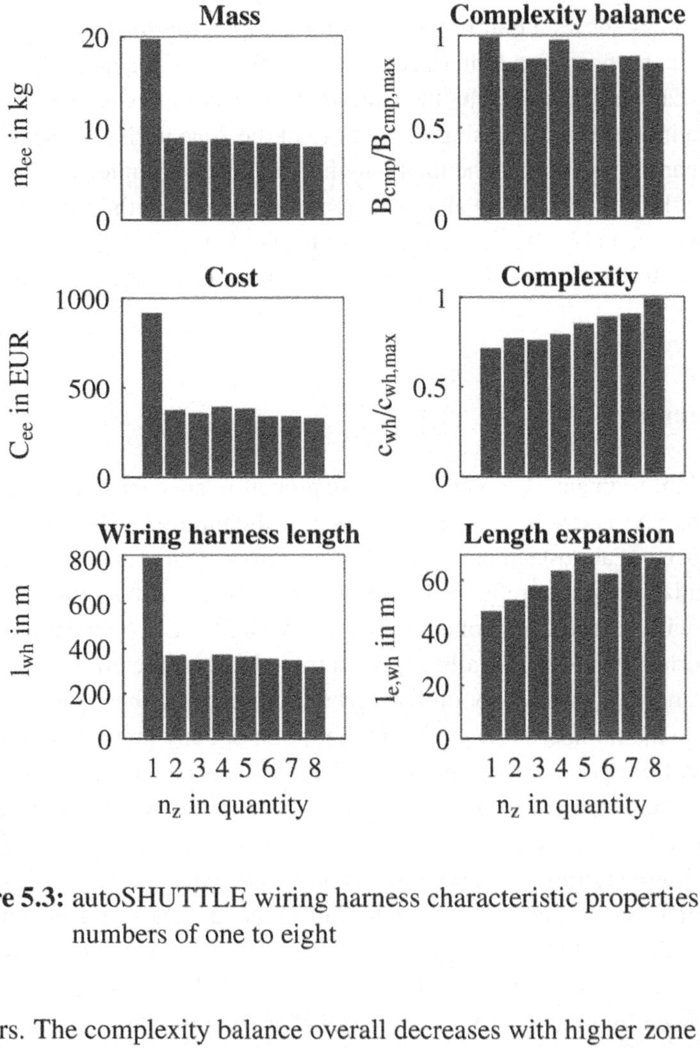

Figure 5.3: autoSHUTTLE wiring harness characteristic properties for zone numbers of one to eight

corners. The complexity balance overall decreases with higher zone numbers, even though three and seven indicate local optima solutions. Complexity rises continuously for increasing zone number, while four and six zones show local minima. Since the complexity of the wiring harness is the sum of the complexity of each zonal wiring harness, local minima indicate highly suitable combinations of components and interfaces. The length expansion increases

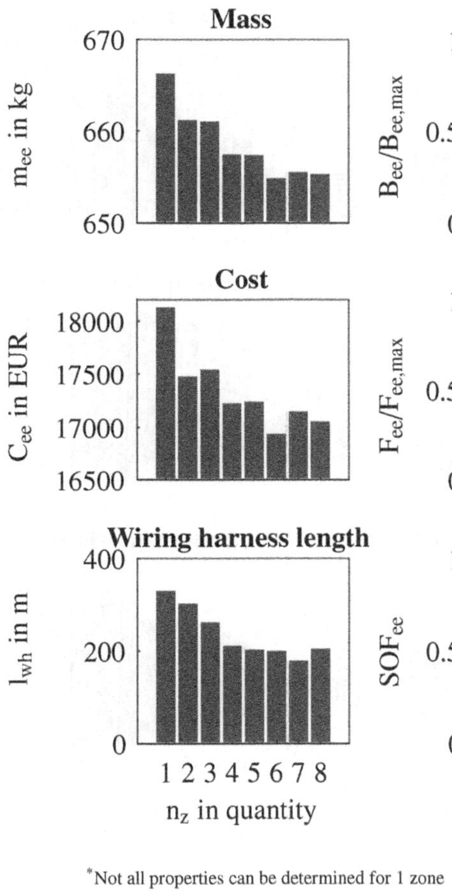

Figure 5.4: Premium-segment SUV EEA characteristic properties for zone numbers of one to eight

with rising zone number, but there is a rapid growth from five to six zones, which could be a result of a strong change in the clustering starting positions.

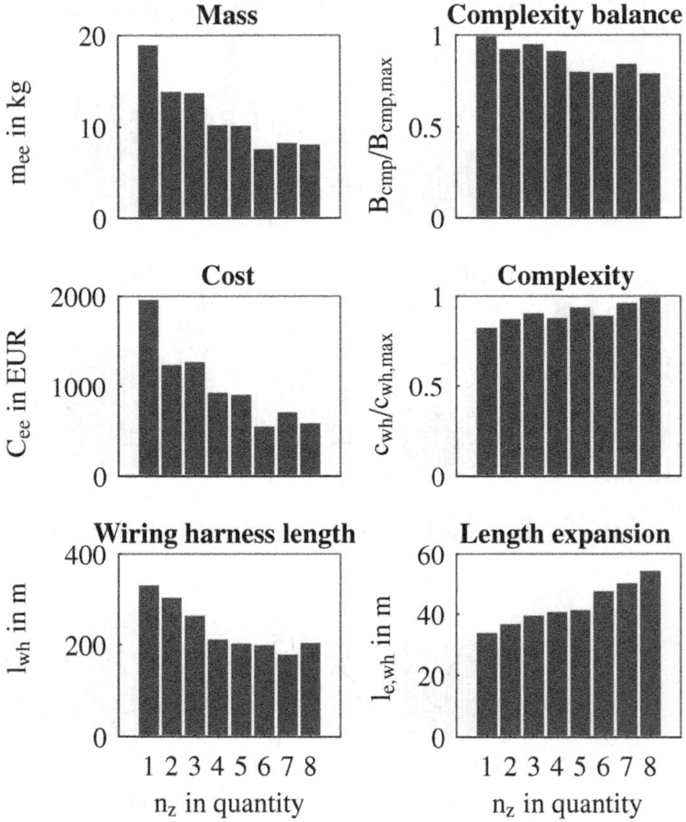

Figure 5.5: Premium-segment SUV wiring harness characteristic properties for zone numbers of one to eight

5.2.3 Comparison

Both simulated vehicles aim for completely different use cases, thus a comparison of the absolute physical values is not reasonable. However, the trends and relation of the properties indicate vehicle specific characteristics, which are meaningful for a comparison. The core properties of both vehicles' EEAs show divergent trends. The autoSHUTTLE demonstrates a strong reduction of all values starting from two, while the use of further zones lead to a minor reduction. As each additional zone comes with extra cost, it is meaningful to choose the lowest zone number, which fulfills the functional and physical requirements. The core attributes of the SUV show a different dependence as wiring harness length and mass reduce with rising zone number. This is an indicator, that the method finds the optimal solution with respect to the individual geometric data and therefore fulfills the criteria of generality. Looking at the whole dependency, the overall costs decrease while there are slight variations and the minimum is reached for six zones. As the wiring harness length starts to rise again for a number of eight, mass and cost likewise increase. Thus a suitable zone quantity may be seven or eight. Comparing the electrical balance of both vehicles, a similar trend can be recognized with higher absolute values for the shuttle. Flexibility also highlights a comparable trend, albeit the SUV shows low values for four, five and seven zones. These differences in the trend might be caused by the variation in the ratio of passenger cabin to vehicle geometry. The SOF is independent of the vehicles geometrical data and hence shows a similar course, with there are minor deviations in the absolute values for two to four zones. The relations and trend of the physical values as well as the flexibility highlight dependencies on the vehicle geometrical properties, consequently both differ between the two vehicles.

While a discussion of the entire EEA attributes is essential, the wiring harness properties provide specific information that needs to be considered before coming to a decision. The core properties of the autoSHUTTLE's wiring harness experience heavy changes when a second zone is introduced. Further zones lead to minor reduction, while the minimal value for all of them is reached at a quantity of eight. The SUV's properties undergo a more gradual change, especial with respect to the wiring harness length. Although cost and mass follow this trend, there are pairs of almost similar values. The complexity balance of the shuttle shows little deviation in the range of two to eight zones with one exception at four zones. The nearly equal component distribution in all four vehicle corners results in a high complexity balance. The shuttle does not have such an exception, but the values in the range of two to four are around 0.1 higher than for greater zone numbers. The complexity for both vehicles' wiring harnesses increases with the zone number. The change in the shuttle's properties is more gradual and the SUV's values have local minima at a quantity of four and six. For the length expansion it is the other way around, the SUV shows a more gradual gradient, whereas the shuttle's values rise until they reach the maximum for a quantity of five and are almost constant thereafter with an local minimum at six zones. The shuttle's wiring harness values provide motivation to use the highest possible zone number, while the local optima in complexity balance or length expansion may influence the zone quantity with respect to the architectures requirements. The SUV wiring harness cost and mass lead to a recommendation of six zones, length and its expansion indicate five zones as an optimum, while the functional properties disclose advantages for a zone number of three.

Both architecture and wiring harness properties embody meaningful variants, but they do not match in some points. The most suitable variant can only be identified when the attributes of both are overlapped and gauged with the requirements. While it is evident that the vehicle packaging has an influence on the results, the extent is unclear. Consequently, in the following sections modifications are applied to the automobiles. At first the powertrain is changed as its cables are relatively heavy and thus may have a major influence on the wiring harness. Second, the door concept of the vehicle is varied to identify the influence of a geometry shift on the wiring harness routing, resulting effects on the properties and consequences on the entire architecture.

5.3 Powertrain Concepts Variation

The powertrain has the highest power consumption of all electronic components in an EV. To identify the interdependence of powertrain and zonal EEA, only the drive components are changed. Based on common PTs in EVs, the following four variants are chosen:

- P1: 4 wheel hub drives (4 WD)
- P2: 2 axle drives (2 AD)
- P3: 1 front axle drive (1AD: F)

- P4: 1 rear axle drive (1AD: R)

The simulations in this chapter are conducted to the SUV, as it is more similar to a series automobile. The two axle drive configuration is most common, but the other variants are applicable and may be more advantageous. Figure 5.6 shows the effects on the architecture level. Analyzing the results on the architecture level, one can recognize, that the weight diverges at low zone numbers with especially P4 as worst variant, but for higher zone numbers the weight is almost equal. This behavior can be explained through the overall reducing wiring harness length and therefore improved routing by increasing zone numbers. The electrical balance shows for the variants P1, P2 and P3 at all zone numbers similar results of 0.7, while P4 performs significantly worse. The SOF rises with increasing zone numbers for the variants P1, P2 and P4, while P3 shows especially at zone numbers up to four low values. The differences in electrical balance and SOF are caused by the unsymmetrical power distribution through the vehicle geometry due to a single driven axis.

Figure 5.7 visualizes the core properties related to the wiring harness. The combination of P4 and low zone numbers can clearly be identified as an expensive variant that should be avoided. In contrast P2 shows for all zone numbers the best results for these properties and is therefore an appropriate variant. This is caused by the fact, that the vehicle was initially designed for this variant. Discussing the global trends of the three properties one can recognize, that the values decrease until a zone number of six or seven is reached. This confirms the selection of the zone numbers of one to eight. However it must be said, that there are specific solutions that do not follow this trends and show good properties for a specific combination of zone number and variant, such as the cost for P3 and three zones. This behavior points out the necessity for not only deriving trends, but analyzing all combinations as intended by this method. The structural properties of the wiring harness are shown in Figure 5.8. The total complexity rises due to the simple addition of the individual complexity values. The balance of the complexity on the other hand shows a diverse behavior with better results in the range of four to six zones, indicating that a zone number in this range will result in the most consistent complexity spread over the zones. This is caused by a even component to ZCU alignment. The length expansion rises by gaining zone numbers due to the fact, that separate sub-harnesses are

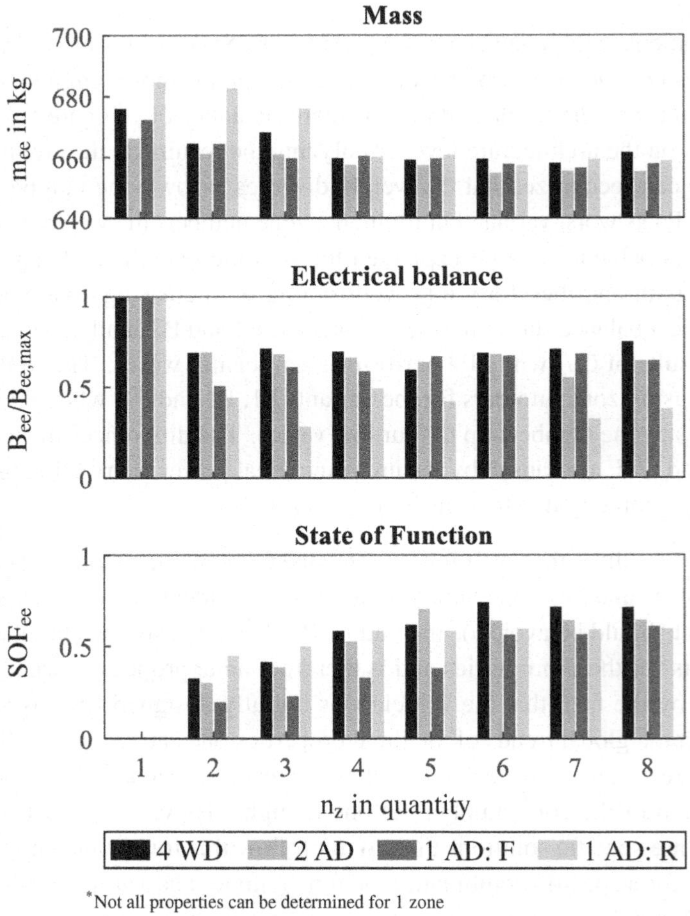

Figure 5.6: Architecture properties of the powertrain concepts P1 to P4

created. However P2 as well as four zone indicate good solutions with respect to the length expansion.

Summarizing these three graphs one can identify the influence of the powertrain on the properties of the EEA and the wiring harness. On architecture level, P1 and P2 show a higher electrical balance, especially for low zone quantity and

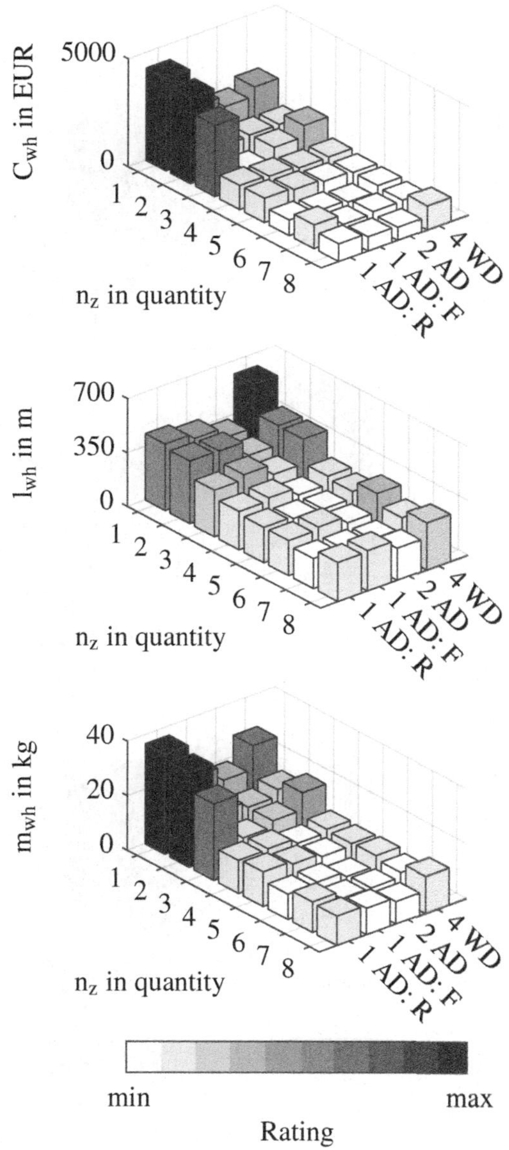

Figure 5.7: Wiring harness core properties of the powertrain concepts P1 to P4

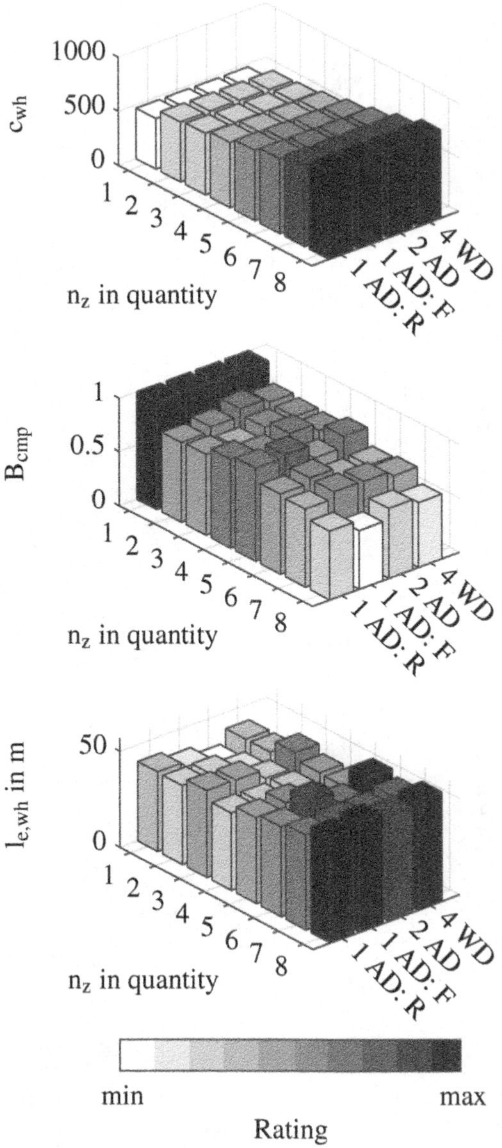

Figure 5.8: Wiring harness structure properties of the powertrain concepts P1 to P4

their SOF is clearly greater independent of the zone number. Looking at the wiring harness, most properties are influenced by the powertrain variants, albeit mass and cost are significantly affected, whereas P2 shows the best results. All in all it can be said, that the variant P2 and zone numbers in the range of four to six are the most promising. The identification of the best variant in this selection can only be determined with respect to the specific requirements.

5.4 Geometric Design Modifications

Geometric modifications are often needed in early, but also in later, phases of vehicle development due to changes of other systems. Therefore, an architecture must not be highly sensitive to small and medium changes, but it is nearly impossible to avoid vulnerability to major changes. This section contains an investigation of two aspects. On the one hand it must be evaluated how sensitive the zonal architecture is to changes, on the other hand it is necessary to identify in which way geometric modifications also influence them. In this section five additional variants are evaluated, which correspond to small and medium changes. The first letter in the brackets indicates the position of the door: R for right, L for left and B for both sides. The second expression differentiates between FH for full size and SH for short size of the door.

- D1: Door on the right side, glass from 1 m to 2 m in height (R/FH)
- D2: Door on the right side, glass from 0.3 m to 2 m in height (R/SH)
- D3: Door on the left side, glass from 1 m to 2 m in height (L/FH)
- D4: Door on the left side, glass from 0.3 m to 2 m in height (L/SH)
- D5: Door on the both sides, glass from 1 m to 2 m in height (B/FH)
- D6: Door on the both sides, glass from 0.3 m to 2 m in height (B/SH)

The six variants vary in the position and height of the door. As the glass of the doors is translucent and they are movable, it is not possible to place cables behind it, thus it is a limitation for the geometric model and the wiring harness routing. Figure 5.9 and Figure 5.10 visualize the results of this simulation study. As this geometric modifications result in minor changes, the figures are limited to the wiring harness, as the effect on the architecture level is relatively small. The core properties comprise cost, length, mass and they are visualized by Figure 5.9. The values for one zone are significantly higher than for the other zones due to the same reasons described in Section 5.2, therefore they are only depicted for the sake of completeness and are not discussed in detail.

All values for zone numbers of two to eight are on a comparable level and indicate minor differences. Analyzing the zone numbers it can be identified, that the minima is at three and eight zones and the maxima is at four and five zones. These extreme can be explained by the combination of zone structure and the corresponding cable paths. When the EEA comprises four to five zones, the centroids tend to the four corners of the vehicle resulting in wire routes along the longitudinal axis near the vehicle's boundary. Those are highly affected by the door modification and consequently their routed path increases. For the case that both doors are fully blocked (D5), these connections are bundled and routed along the vehicles longitudinal axis, resulting in an reduction of the length expansion. This effect indicates the influence of the bundling coefficient, which can be adjusted to solve even such problems. By comparing the different door concepts one can identify, that for most zone numbers the variants D3 to D6 perform better than D1 and D2. The difference between the short height (/SH) and full height (/FH) variants are minor and thus can not be distinguished in the figure. Additionally, there is a minimum for D3 and D4 in combination with three zones. These trends are partly reflected by the structural properties. An increase in complexity is normal due to the addition of the individual complexity. Hence a good variant is characterized by an under average increase or a slight decrease, which is embodied by the variants D4 and D6. The same applies to the complexity balance, which shows low values for eight and a minima for six zones. The expansion length of the wiring harness demonstrates a constant increase for zone quantities up to five, a minima at six zones and only a slight gain for seven and eight zones compared to the former trend. Overall, the variants D3 and D4 show marginally better results for the complexity balance and the length expansion, albeit there are other variant optima such as the electric balance of D1 for seven zones or of D6 for five zones.

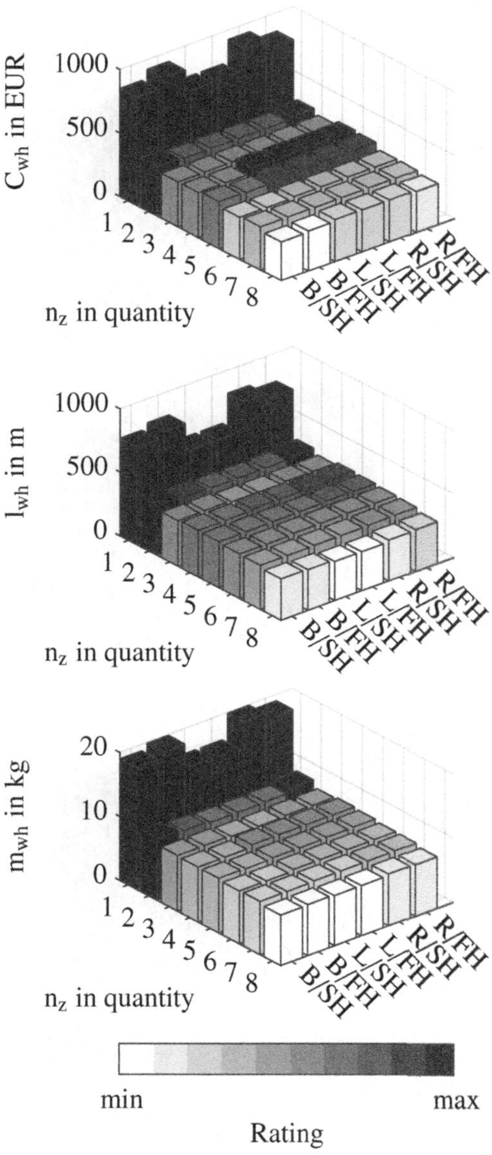

Figure 5.9: Wiring harness core properties for the door concepts D1 to D6

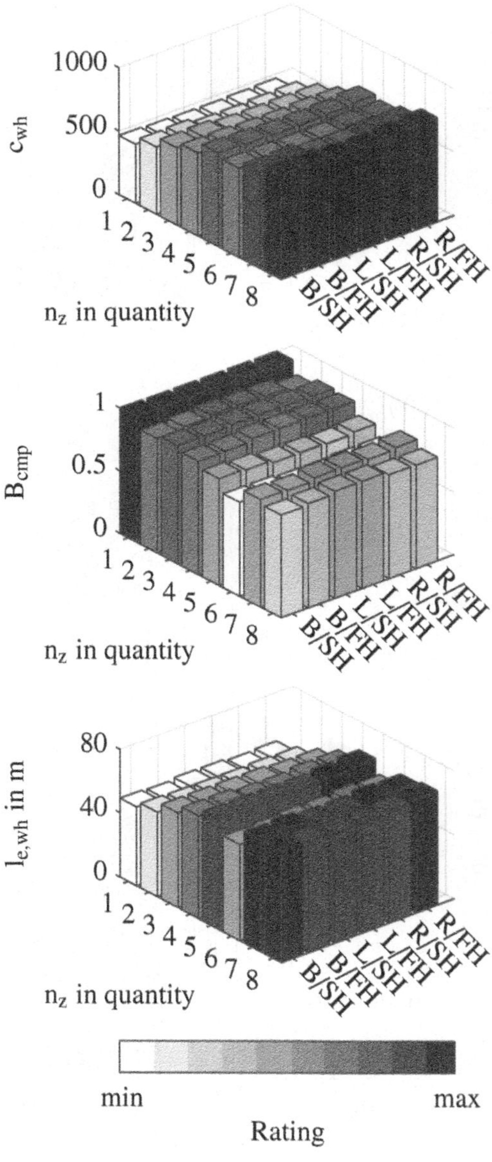

Figure 5.10: Wiring harness structure properties for the door concepts D1 to D6

5.5 Battery Number Variation

As the goal of zonal EEA usage is the simplification of architectures, new battery concepts with respect to spatial accessibility can play a role. Thus the following four battery concepts are evaluated in this section:

- B1: One traction battery/-connector (1 Bat.)
- B2: Two traction batteries/-connectors (2 Bat.)
- B3: Four traction batteries/-connectors (4 Bat.)
- B4: n_z traction batteries/-connectors ($\mathrm{n_z}$ Bat.)

Today most vehicles come with one traction battery that is accessible by one connector. In addition to the established variant B1, there are B2 to B4 that are not integrated in series vehicles today. B2 has been discussed in recent years as automated driving will remove the driver as the fallback solution for the case of failures, hence reliability must be increased [182, 183]. In the case of the energy storage this can only be realized by integrating a second one. B3 is realized in the UNICARagil platform using four traction batteries, albeit this is not meaningful for series vehicles due to the associated increase in effort needed for the development and production. Thus B3 should be implemented using four connectors and not more than two batteries. The same applies to B4, albeit it is meaningful to increase the number of battery connectors to the quantity of zones to enable short transmission paths. For a better understanding, the variants in this section are described by their connector number, albeit for the case for B2 multiple batteries are possible.

Figure 5.11 and Figure 5.12 show the core and structure properties of the four variants. Cost and mass decrease with higher connector quantity and with exceptions rise with the zone number. This trend is also present in the length of the wiring harness, albeit there are some maxima for zones numbers of three, six and eight in combination with B1 to B3. This occurs due to the fact, that for all variants of B1 to B3 identical battery connector positions were used for a fair evaluation.

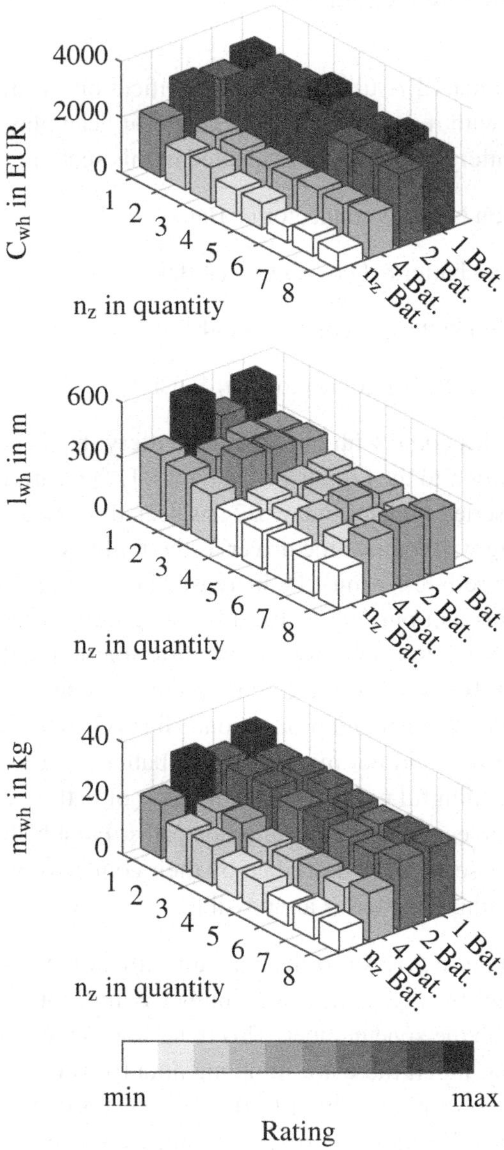

Figure 5.11: Wiring harness core properties for the battery concepts B1 to B4

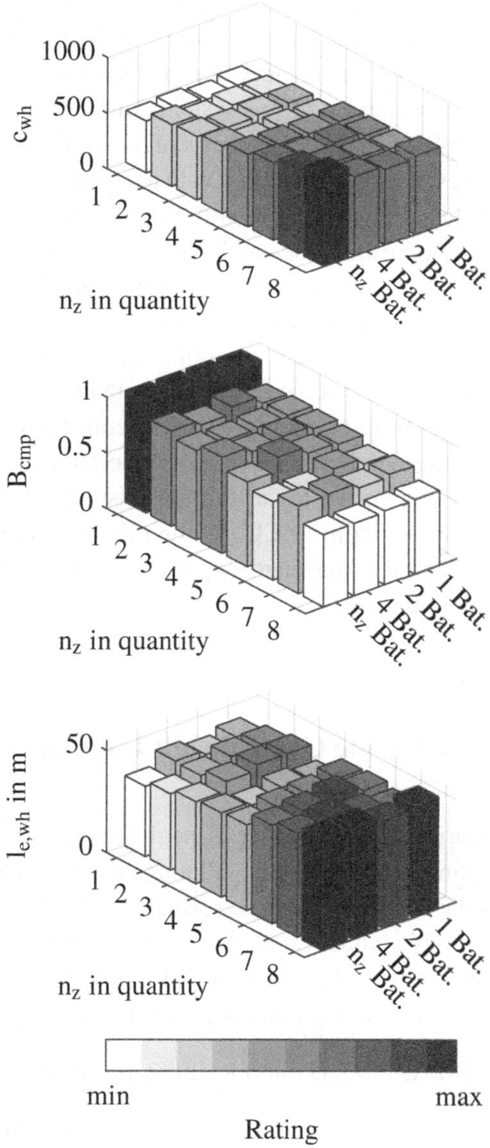

Figure 5.12: Wiring harness structure properties for the battery concepts B1 to B4

These combinations do not match properly with respect to the harness length, nevertheless their cost and mass follow the trend of the entire data set. Looking at the complexity value, one can identify that the increase of zone and connector quantity drives the complexity of the systems due to higher branching and expansion. The complexity balance reaches a high number for similar results in the range of two to seven zones in addition with a few minima and maxima. This indicates, that the clustering process identifies good results for a diverse set of parameter; moreover, there are some combinations that show a slight divergence due to the matching of the cluster and battery positions. The length expansion increases with rising zone numbers and decreasing connector quantity, although in specific cases a reduced number of connectors results in less expansion. Consequently it can be said, that there is a high potential to simplify the wiring harness in zonal EEAs through multiple battery connectors. The best results are achieved when the numbers of battery connectors and zones are equal. In addition it should be strove for a minimal distance between the battery connector and its corresponding ZCU. While the total complexity rises, the complexity balance as well as the length expansion can be reduced by increasing the numbers of battery connectors.

5.6 Unsymmetrical Zone Control Unit Interfaces

One of the boundaries defined earlier is neglected in order to compare its influence on the optimization approach. Although the identical number of ZCU interfaces reduces the component cost, it severely limits the optimization of the structure. To identify this influence, the limitation for the interfaces is removed in the course of this section. Figure 5.13 and Figure 5.14 show the core and structure properties of the wiring harness based on the following four variants:

- Z1: autoSHUTTLE with ZCU interface limitations (SHU)
- Z2: autoSHUTTLE without ZCU interface limitations (SHU/imb)
- Z3: SUV with ZCU interface limitations (SUV)
- Z4: SUV without ZCU interface limitations (SUV/imb)

By focusing on the core properties first it can be identified that for the imbalanced variant Z4 these characteristics achieve lower values than Z3 for two and three zones. Besides these differences, there are minor divergences between most variants, indicating that this boundary condition influences the clustering procedure only slightly. However, some variants are advantageous when the interface number is not balanced, due to higher flexibility in the clustering process and therefore the improvement of an individual aspect such as mass. The structural properties show more diverse effects and thus are discussed separately. All complexity values follow a common trend, albeit the imbalanced variants show noteworthy reductions for several zone numbers such as three and eight for the SUV and all zone numbers for the autoSHUTTLE. The length expansions indicates a similar effect, albeit there are some variants where the balanced variants Z1 and Z3 achieve better results. The complexity balance values are conclusive to the other core and structure properties. The divergence of the variants without interface limitations from the regular ones are for most cases up to 30 %, however there are solutions that show an outstanding difference of around 80 %.

Summarizing the results of both property sets it can be said, that the clustering process is influenced slightly by removing interface limitations, due to minor deviation of core and structure properties. However, in particular, the complexity balance profits from the evenly distributed ZCU interfaces. Consequently, it can be said, that balancing the components per ZCU enables modularization and therefore cost reduction for the same, but may result in structural drawback. To decide whether it is beneficial to balance the component allocation, it is necessary to weight up the effort for wiring harness as well as ZCU development of the individual scenario.

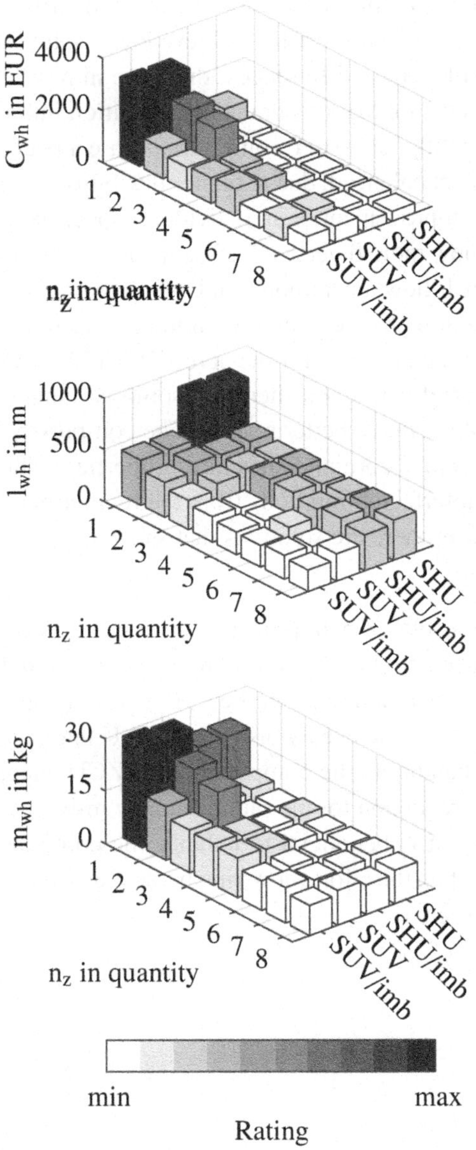

Figure 5.13: Wiring harness basic properties for the concepts Z1 to Z4

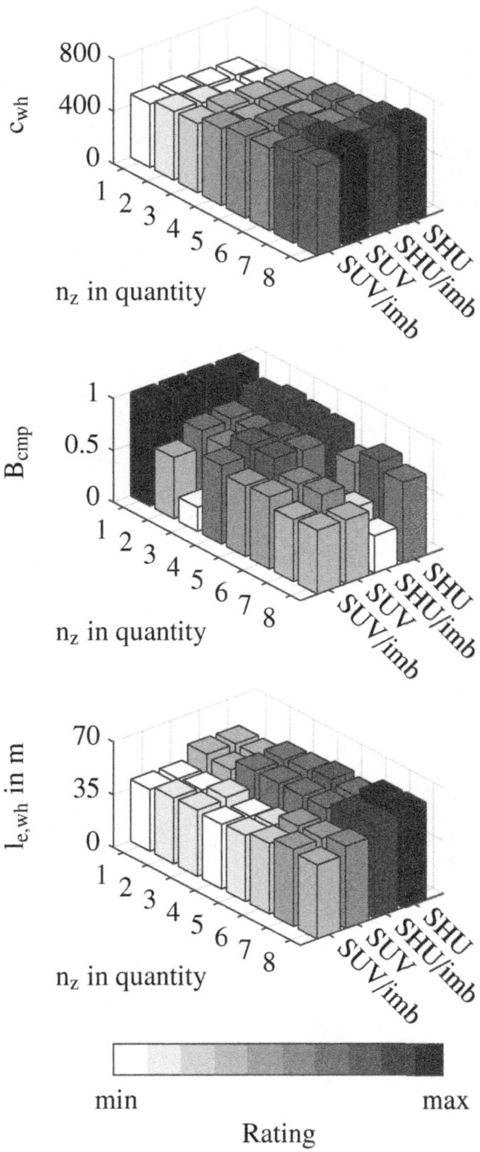

Figure 5.14: Wiring harness structure properties for the concepts Z1 to Z4

6 Summary and Outlook

The research questions of this thesis were derived from the structural shift in EEAs. This is linked to the fact, that decentralized or domain centralized architectures were structured by technological or functional aspects, while a zone centralization requires the structuring by geometrical aspects. As the problem posed involves an interdisciplinary complex system, a limitation to relevant aspects is essential for the identification of a meaningful solution. Vehicle software development has also shifted in the recent years towards higher flexibility by means of service-oriented software approaches that are independent from the hardware, and can have an addition and update of functionalities along the entire product life cycle [184]. This enables the independent design of the hardware platform, with the requirements of the software architecture as boundaries. To facilitate a learning effect based on the research actions taken in the course of this dissertation, a critical reflection is conducted on the basis of the research questions. Thus the capabilities and drawbacks as well as remaining open questions are discussed. To ensure that other disciplines than automotive engineering can profit from this dissertation, an outlook is given about the other applications that this method can be applied to.

Addressing the first question a method was developed that can be applied to the early development phases of vehicles, where architectures are defined and refined. Knowledge about interdependence of systems, especially of the EEA with electric components and vehicle packaging are invaluable. The method determines clusters in the spatial dispersion of electrical components, comprising the identification of cluster centroids and the component alignment. Based on the clusters, the traction battery and the low-voltage powernet are designed. The simplification of the wiring harness by weight and length reduction as well as the separation into individual zone wiring harnesses is one of the major potentials of the zonal EEAs approach. To utilize the potentials the design of the wiring harness complements this holistic design approach embracing requirement supported optimal path identification and cable bundling. Comparing the used approaches to the state of the art, one needs to consider that

J. Maier, *Optimization of Zonal Electric/Electronic Architectures Using a Coupled Modeling Approach of Package and Wiring Harness*,
Wissenschaftliche Reihe Fahrzeugtechnik Universität Stuttgart,
https://doi.org/10.1007/978-3-658-48188-9_6

the problem discussed is highly product driven and not by science. As the automotive industry has not introduced a zonal EEA in any series vehicles, there are only few approaches dealing with zonal architectures, but none with these specific aspects. k-means clustering is largely used in spatial clustering, such as by Jayanti, but not in combination with the boundaries specified in this thesis [185]. Traction battery design in the context of automotive engineering using genetic algorithm has been applied by several authors before, but without considering the integration into vehicle package [13, 186]. Path optimization in graphs using the Dijkstra algorithm has been intensively researched, and for this work extended in the context of wiring harness optimization with recent findings [151, 177]. Consequently, existing approaches are adapted to zonal EEAs, improved and aggregated to form a holistic approach, that is able to optimize the entire architecture instead of individual aspects.

The answer for the second research question is closely connected with the first one. To enable the integration of an obtained solution, it is necessary to model all relevant aspects in a common geometry model covering the zone allocation and segmentation, the placement in the vehicle package as well as the path routing of cables. Therefore, a three-dimensional based graph is meshed with a voxel representation of the vehicles geometry to ensure end-to-end compatibility with the optimization method.

The last question is a premise for answering the first one, as the wiring harness properties crucially influence the architecture attributes. Complexity is an intangible attribute and is highly dependent on the system and the environment it is exposed to, therefore its determination is complicated and individual. The aim of this thesis is to answer this question by two approaches. The first was a mathematical approach defined by Sinha, which focuses on complexity of the topology and is converted to mathematical graphs [77]. The benefit of this method is the pure evaluation using mathematics relation, thereby avoiding any subjective influence by the person assessing. The downside of this approach is, that it is an attempt to make complexity tangible, but will never be able to handle all aspects that embody the complexity of a system. The second approach is the collective evaluation of critical attributes and their combination. Both approaches are applied and their results are discussed, however they are not coupled. The separate application enables a comparison to be made of the complexity of different variants and therefore answers this question. The possible coupling of these two perspectives can be beneficial if the output is an absolute measure comprising both approaches.

6.1 Research Goals and Contribution

According to Hogenmüller, 80 % or more of the OEMs will switch to zonal EEAs. Consequently there will be a need for methods to design and integrate zonal EEAs [187]. It is crucial to be aware, that there is no optimal EEA for all use cases, vehicles or platforms. When designing a new platform, flexibility or a specific safety level might be highly important, while the aim of the development of a new variant is related to the refinement of cost or mass. Hence it is not meaningful to evaluate all functions in a combined approach using a cost functional, instead promising variants should be simulated and the system properties evaluated with respect to the requirements. As the wiring harness simplification is one of the major potentials of zonal EEAs, the difference on architecture level for properties such as mass or cost may be a few percent. The functional properties on architecture level as well as most characteristics of the wiring harness show significant differences when the zone number is varied. Geometric as well as powertrain modification also leads to minor deviations showing advantages and disadvantages. Based on the simulation results, the solution space is reduced by selecting the overall best variants. The method does not support the balancing of properties with respect to the system distinct requirement, albeit these can be considered in individual steps of the approach. Consequently, this method will not fully substitute the analytical skills and knowledge of an engineer necessary for the design of boundary conditions and the evaluation of the individual results with respect to the system requirements. However, the results can be concentrated in a design structure matrix [188]. Figure 6.1 shows an example of a reduced design structure matrix with equal metric weighting. It is reduced to the combinations discussed in the simulation, therefore the rows indicate the zone numbers of one to eight and the columns the modifications. It enables the objective comparison of variants with respect to their quality and by this supports design decisions. By variation of the metric weighting, the significance of the individual metrics can be varied and by this user-oriented solutions are designed.

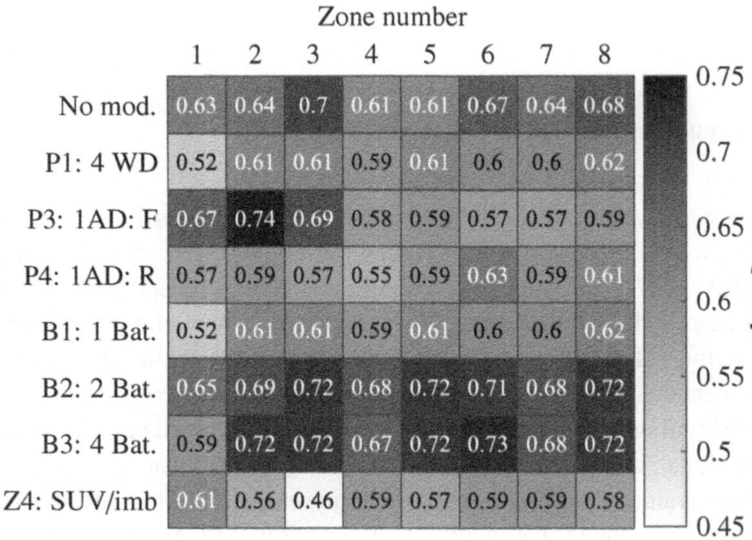

Figure 6.1: Reduced design structure matrix of the Premium SUV with equal metric weighting

Besides this methodical drawback, there are some technical simplifications that leave potential for further improvement. The geometrical model is based on a discrete approximation of the vehicles geometry, thereby causing inaccuracies in the voxel reservation of components and in the vehicles contours. This can lead to unexploited space in the vehicle, because a voxel is reserved whenever a component projects into it, albeit it might not occupy the whole. A second downside of the approximation is the pointlessness of realizing a collision control between components and wiring harness, as the voxel size is significantly larger than the cable diameter.

The zone model is developed using various references, but there are major gaps in the state of knowledge today. Therefore, assumptions include the idea that a similar amount of components respectively interfaces per ZCU is reasonable and that they are expandable in terms of computing capability as well as data and electric power transmission. Furthermore is the number of zones limited to eight, but as some properties reach a minimum at the maximum zone number, an increase of the same could result in further improvements. Albeit this presumptions seem reasonable today, it is not clear how exactly the design of a ZCU will look like in a series vehicle. This also applies to the price assumption for ZCUs, although it is based on an estimation made by professionals [10]. Function and software architecture were completely neglected in the optimization process and considered in terms of boundary conditions such as interface number and transmission technology. The purpose is to develop hardware and software with a more loose connection to each other, as all potentials of zonal EEAs can only utilized by a more generic hardware design and service oriented software architecture [189, 190]. Therefore, the transmission technology is designed considering a worst case scenario, which might result in overdimensioned data transmission.

There is one further aspect which needs to be mentioned when discussing the validity of the results. The cross section of single wires as well as bundles need to be designed dependent on the dynamic profile of current and temperature, which were not considered in this dissertation. The cross section of a cable is selected with respect to the maximum current and therefore is suitable in general, albeit a section with a high current density flow can lead to hot spots.

6.2 Outlook

The method introduced in this dissertation is developed and investigated with a focus on specific systems and applications. Beyond that, it is meaningful to discuss its interaction with the present systems and how a replacement can take place. Additionally, its transferability is exemplified by a reduction of the investigated system to the essential properties and identification of other systems possessing these characteristics.

6.2.1 Harmonization with Established Architectures

Zonal EEAs are often presented as a way out of the complexity of todays vehicles architectures. Although their potential can not be denied, design and integration of such an approach poses various challenges. While this dissertation focuses on the design and integration of hardware with respect to the package, there are still various challenges for hardware and software development. One aspect that affects all disciplines and layers is the harmonization with established architectures, respectively domain-centralized architectures. Wang et al. discuss recent trends in academia and industry and propose two future development directions [26]. The first is the continuation of the domain-centralized approach by functional centralization in the domains respectively DCUs and in subsequent steps the consolidation of different domains to further reduce the number of components in particular ECUs and the wiring harness. This approach will finally lead to the consolidation of most functionalities in a central computing unit with few DCUs. The second development direction is the zone centralization leading to ZCUs with particular high functionality. In the subsequent steps of this approach a central computing unit will be introduced and by this the functional range of the ZCUs will reduce.

OEMs are bound to their product cycle, which is around six to seven years for passenger cars and in many cases there is a product upgrade after around three years. In the past, short development time and the immense effort needed to fully replace an existing EEA has resulted in a gradual replacement [191]. As development time shortens and system complexity rises, an incremental architecture shift is expected. One option could be the integration of zonal architectures and consolidation of the ECUs having a small functional range with little interdependence and temporary retention of the highly complex ECUs.

6.2.2 Transferability

The method proposed in this thesis enables the design and integration of the hardware layer for zonal EEA in vehicles. It is designed for a specific problem, but generalization and transfer to other applications are possible. Therefore, the problem should be discussed more analytically. From an analytical viewpoint, the problem can be reduced to the identification of meaningful clusters for electronic components in a three-dimensional environment and their connections, both under application specific requirements. Consequently, the problem is applicable to most mechatronic systems, albeit the boundary conditions differ by the utilization. However, two further criteria need to be fulfilled for a meaningful usage. These criteria correlate with the reasons for the shift towards zonal EEA in the automotive sector. The large number of electronic components requiring energy and data transmission in combination with their intensive interlinkage result in a complex and inefficient wiring harness that makes an architecture shift vital. For less complex systems, which have either fewer components or less dependencies, zone centralization of the architecture might not be meaningful. The second criteria concerns the spatial dispersion of the components. The method is most beneficial when the distance between the components is high, the nesting of the structure is complicated and there are so many interconnections that their combination results in a wiring harness that is so complex and costly that its simplification is inevitable. Therefore, the method can be extended to other vehicles as well as, but not exclusively, to the following. Cui et al. introduced the zonal EEA in a humanoid robot to decrease cable length and mass [192]. Sulligoi discusses zonal electrical distribution systems for ships and oil platforms, which both satisfy the above criteria [193]. Alexander mentions that an established variant of electric power system for aircraft follows a zonal approach [194]. Not only do these three exemplary systems fulfill both criteria, furthermore there are already concepts or implementations of a zonal EEA in them. In conclusion, the developed method can be used used to design and evaluate the hardware layer of zonal EEAs in vehicles, and an application on other CPS is, with adaption of boundary conditions, possible.

Bibliography

[1] Aptiv plc. *Evolution of Vehicle Architecture: [Article]*. 2018. URL: https://www.aptiv.com/en/insights/article/evolution-of-vehicle-architecture (visited on 03/01/2024).

[2] Ross McQuat. *Cars are Made of Code: [Article]*. Ed. by NXP Semiconductors N.V. 2017. URL: https://www.nxp.com/company/blog/cars-are-made-of-code:BL-CARS-MADE-CODE (visited on 03/01/2024).

[3] World Resources Institute, ed. *Climate Watch Historical GHG Emissions*. Washington, DC, 2022. URL: https://www.climatewatchdata.org/ghg-emissions.

[4] Joao P. Trovao. "Automotive Electronics Market Evolution [Automotive Electronics]". In: *IEEE Vehicular Technology Magazine* 15.1 (2020), pp. 107–118. ISSN: 1556-6072. DOI: 10.1109/MVT.2019.2954734.

[5] Ondrej Burkacky et al. *Rewiring car electronics and software architecture for the Roaring 2020s*. Ed. by McKinsey Center for Future Mobility. 2021.

[6] Ondrej Burkacky, Johannes Deichman and Jan Paul Stein. *Automotive software and electronics 2030: Mapping the sector's future landscape*. Ed. by McKinsey & Company, Inc. 2019.

[7] Laura Sophie Brandt. "Architekturgesteuerte Elektrik/Elektronik Baukastenentwicklung im Automobil". Doctoral Dissertation. Munich, Germany: Technical University of Munich, 2016.

[8] Arturo Angel Lara Rivero. "From complex mechanical system to complex electronic system: the case of automobiles". In: *International Journal of Automotive Technology and Management* 14.1 (2014), p. 65. ISSN: 1470-9511. DOI: 10.1504/IJATM.2014.058366.

[9] Joao P. Trovao. "The Vehicle Industry Is Moving Fast [Automotive Electronics]". In: *IEEE Vehicular Technology Magazine* 17.1 (2022), pp. 98–107. ISSN: 1556-6072. DOI: 10.1109/MVT.2021.3130399.

J. Maier, *Optimization of Zonal Electric/Electronic Architectures Using a Coupled Modeling Approach of Package and Wiring Harness*, Wissenschaftliche Reihe Fahrzeugtechnik Universität Stuttgart, https://doi.org/10.1007/978-3-658-48188-9

[10] Ondrej Burkacky et al. *Getting ready for next-generation E/E architecture with zonal compute: The next generation of electrical/electronic architecture will play a vital role in the software-defined vehicles of the future—OEMs and chip makers need to be strategic in their choices.* Ed. by McKinsey & Company. 2023. URL: https://www.mckinsey.com/industries/semiconductors/our-insights/getting-ready-for-next-generation-ee-architecture-with-zonal-compute.

[11] Leoni AG. "Leoni uses zonal architectures to reinvent the wiring system: [Press release]". In: (2023).

[12] Stefan Caba and Gerhard Becker. "Modularization and E/E Architecture for Greater Sustainability". In: 125 (2023), pp. 18–23. DOI: 10.1007/s38311-023-1487-x.

[13] Jan Nicolas Becker. *Dimensioning and Optimization of Hybrid Li-Ion Battery Systems with different Optimization Goals*. RWTH Aachen University, 2017. DOI: 10.18154/RWTH-2017-09278.

[14] Felix Frank. "Optimierter Hochvoltbatterieentwurf hinsichtlich mechanischer, thermischer und elektrischer Randbedingungen unter Berücksichtigung der Produktionskosten". PhD thesis. DuEPublico: Duisburg-Essen Publications online, University of Duisburg-Essen, Germany, 2019. DOI: 10.17185/DUEPUBLICO/70278.

[15] Kurt Lewin. "Action Research and Minority Problems". In: *Journal of Social Issues* 2.4 (1946), pp. 34–46. ISSN: 0022-4537. DOI: 10.1111/j.1540-4560.1946.tb02295.x.

[16] Lucienne T.M. Blessing and Amaresh Chakrabarti. *DRM, a Design Research Methodology*. London: Springer London, 2009. ISBN: 978-1-84882-587-1. DOI: 10.1007/978-1-84882-587-1.

[17] Miroslaw Staron. *Action Research in Software Engineering*. Cham: Springer International Publishing, 2020. ISBN: 978-3-030-32609-8. DOI: 10.1007/978-3-030-32610-4.

[18] J. Bach, S. Otten and E. Sax. "Classification of Automotive Electric/Electronic Features and the Consequent Hierarchization of the Logical System Architecture". In: *Smart Cities, Green Technologies, and Intelligent Transport Systems*. Ed. by Brian Donnellan et al. Communications in Computer and Information Science. Cham: Springer International Publishing, 2019, pp. 229–255. ISBN: 978-3-030-02906-7. DOI: 10.1007/978-3-030-02907-4_12.

[19] Matthias Kreimeyer and Udo Lindemann. *Complexity Metrics in Engineering Design*. Berlin, Heidelberg: Springer Berlin Heidelberg, 2011. ISBN: 978-3-642-20962-8. DOI: 10.1007/978-3-642-20963-5.

[20] Alessandro Frigerio, Bart Vermeulen and Kees G. W. Goossens. "Automotive Architecture Topologies: Analysis for Safety-Critical Autonomous Vehicle Applications". In: *IEEE Access* 9 (2021), pp. 62837–62846. DOI: 10.1109/ACCESS.2021.3074813.

[21] S. Geldhäuser et al. *Vitesco Technolgies Develops Master Controller for a New Era of Vehicle Motion Control*. Ed. by Vitesco Technologies GmbH. 2021. URL: https://www.vitesco-technologies.com/en-us/press-events/press/tech-day-master-controller (visited on 28/08/2023).

[22] Robert Bosch GmbH. *Vehicle Control Unit: The vehicle control unit as a central E/E architecture component for all powertrain types*. 2019. URL: https://www.bosch-mobility.com/de/loesungen/steuergeraete/vehicle-control-unit/ (visited on 28/08/2023).

[23] Chulsun Park and Sungkwon Park. "Performance Evaluation of Zone-Based In-Vehicle Network Architecture for Autonomous Vehicles". In: *Sensors* 23.2 (2023), p. 669. DOI: 10.3390/s23020669.

[24] Victor Bandur et al. "Making the Case for Centralized Automotive E/E Architectures". In: *IEEE Transactions on Vehicular Technology* 70.2 (2021), pp. 1230–1245. ISSN: 0018-9545. DOI: 10.1109/TVT.2021.3054934.

[25] Angela Gonzalez Marino, Francesc Fons and Juan Manuel Moreno Arostegui. "The Future Roadmap of In-Vehicle Network Processing: A HW-Centric (R-)evolution". In: *IEEE Access* 10 (2022), pp. 69223–69249. DOI: 10.1109/ACCESS.2022.3186708.

[26] Wenwei Wang et al. "Review of Electrical and Electronic Architectures for Autonomous Vehicles: Topologies, Networking and Simulators". In: *Automotive Innovation* 7.1 (2024), pp. 82–101. ISSN: 2096-4250. DOI: 10.1007/s42154-023-00266-9.

[27] Julian Weber. *Automotive Development Processes: Processes for Successful Customer Oriented Vehicle Development*. Berlin, Heidelberg: Springer Berlin Heidelberg, 2009. ISBN: 978-3-642-01252-5. DOI: 10.1007/978-3-642-01253-2.

[28] Lisa Braun. "Modellbasierte Design-Space-Exploration nicht-funktionaler Auslegungskriterien des Fahrzeugenergiebordnetzes". Doctoral Thesis. Karlsruher Institut für Technologie, 2018. DOI: 10.5445/IR/1000081298.

[29] Dusan Kenjic, Marija Antic and Milan Bjelica. "Evaluation of Ethernet Subsystem for Domain Controller in Autonomous Vehicles". In: *2021 Zooming Innovation in Consumer Technologies Conference (ZINC)*. IEEE, 2021, pp. 59–63. ISBN: 978-1-6654-0417-4. DOI: 10.1109/ZINC52049.2021.9499282.

[30] Stefan Mischo et al. "E/E architectures at the crossroads". In: *ATZ Elektron Worldw (ATZ- elektronik worldwide)* 3 (2008), pp. 10–13. DOI: 10.1007/BF03242187.

[31] Christian Schäfer and Rainer Denkelmann. "Sustainable E/E Architecture Power Supply and Data Transmission for Autonomous Driving". In: *ATZelektronik worldwide* 13.6 (2018), pp. 16–21. DOI: 10.1007/s38314-018-0078-x.

[32] Timo Woopen and et al. *UNICARagil - Disruptive Modular Architectures for Agile, Automated Vehicle Concepts*. Aachen, Germany, Oct, 8th - 10th, 2018. DOI: 10.18154/RWTH-2018-229909.

[33] Dorsa Zaheri et al. "Practical Implementation of a Zonal E/E Architecture for Autonomous Vehicles in UNICARagil". In: *ATZelectronics worldwide* 18.1-2 (2023), pp. 8–12. DOI: 10.1007/s38314-023-1436-x.

[34] Marcus Goth, Dan Keilhoff and Hans-Christian Reuss. "Fault tolerant electric energy supply system design for automated electric shuttle bus". In: *20. Internationales Stuttgater Symposium: Band 2*. Ed. by Michael Bargende, Hans-Christian Reuss and Andreas Wagner. 2020, pp. 441–455. ISBN: 978-3-658-30995-4. DOI: 10.1007/978-3-658-30995-4_40.

[35] Jonas Maier and Hans-Christian Reuss. "Handling System Complexity in Zonal E/E Architectures". In: *Transportation Engineering* 13 (2023), p. 100195. ISSN: 2666691X. DOI: 10.1016/j.treng.2023.100195.

[36] Christian Pana et al. "Die neueste Revolution auf der Straße: Auto als Service: The newest Road Revolution: Car as a Service". In: *Automotive meets electronics*. Ed. by Michael Wahl. GMM-Fachbericht. Berlin and Offenbach: VDE Verlag GmbH, 2017, pp. 113–116. ISBN: 3800743698.

[37] Ben Rutten and ROy Cobbenhagen. "Future Trends in Electric Vehicles Enabled by Internet Connectivity, Solar, and Battery Technology". In: *Automotive Systems and Software Engineering*. Ed. by Yanja Dajsuren and Mark van den Brand. Cham: Springer International Publishing, 2019, pp. 323–346. ISBN: 978-3-030-12156-3.

[38] Jamil Hamadneh, Szabolcs Duleba and Domokos Esztergár-Kiss. "Stakeholder viewpoints analysis of the autonomous vehicle industry by using multi-actors multi-criteria analysis". In: *Transport Policy* 126 (2022), pp. 65–84. ISSN: 0967070X. DOI: 10.1016/j.tranpol.2022.07.005.

[39] Carlos Llopis-Albert, Francisco Rubio and Francisco Valero. "Impact of digital transformation on the automotive industry". In: *Technological forecasting and social change* 162 (2021), p. 120343. ISSN: 0040-1625. DOI: 10.1016/j.techfore.2020.120343.

[40] I. Fadhil and C. Shen. *Electric vehicles market monitor for light-duty vehicles: China, Europe, United States, and India, 2022*. Ed. by International Council on Clean Transportation. 2023. URL: https://theicct.org/wp-content/uploads/2023/06/Major-Mkts_briefing_FINAL.pdf.

[41] Piergiuseppe Mallozzi et al. "Autonomous Vehicles: State of the Art, Future Trends, and Challenges". In: *Automotive Systems and Software Engineering*. Ed. by Yanja Dajsuren and Mark van den Brand. Cham: Springer International Publishing, 2019, pp. 347–367. ISBN: 978-3-030-12156-3.

[42] Junghwan Lee and Longda Wang. "A Method for Designing and Analyzing Automotive Software Architecture: A Case Study for an Autonomous Electric Vehicle". In: *2021 International Conference on Computer Engineering and Artificial Intelligence (ICCEAI)*. IEEE, 2021, pp. 20–26. ISBN: 978-1-6654-3960-2. DOI: 10.1109/ICCEAI52939.2021.00004.

[43] Rolf Isermann. *Automotive Control*. Berlin, Heidelberg: Springer Berlin Heidelberg, 2022. ISBN: 978-3-642-39439-3. DOI: 10.1007/978-3-642-39440-9.

[44] Volker Scheuch. "E/E architecture for battery-electric vehicles". In: *ATZelektronik worldwide* 6.6 (2011), pp. 14–19. DOI: 10.1365/s38314-011-0058-x.

[45] Thilo Streichert and Matthias Traub. *Elektrik/Elektronik-Architekturen im Kraftfahrzeug*. Berlin, Heidelberg: Springer Berlin Heidelberg, 2012. ISBN: 978-3-642-25477-2. DOI: 10.1007/978-3-642-25478-9.

[46] Beate Bender and Kilian Gericke. *Pahl/Beitz Konstruktionslehre*. Berlin, Heidelberg: Springer Berlin Heidelberg, 2021. ISBN: 978-3-662-57302-0. DOI: 10.1007/978-3-662-57303-7.

[47] Peter Wallin. *Key Challenges in Decision Making for Automotive E/E Architectures*. Vol. 93. Mälardalen University Press Licentiate Theses. Västerås, Sweden: Mälardalen University Press, 2008. ISBN: 978-91-85485-97-0.

[48] Martin Jaensch. *Modulorientiertes Produktlinien Engineering für den modellbasierten Elektrik/Elektronik-Architekturentwurf*. Vol. 8. Steinbuch Series on Advances in Information Technology. Karlsruhe, Germany: KIT Scientific Publishing, 2012. ISBN: 978-3-86644-863-6.

[49] International Organization for Standardization. *ISO 26262:2018: Road vehicles - Functional safety*. 2018-12-01.

[50] Andreas Lübke. *Systematischer Bordnetzentwurf: Optimierung der Bordnetzarchitektur mit Hilfe von genetischen Algorithmen: Zugl.: Dresden, Techn. Univ., Diss., 1999*. 1. Aufl. Aachen: Mainz, 1999. ISBN: 3896535609.

[51] Kirsten Matheus and Michael Kaindl. *Automotive High Speed Communication Technologies: SerDes and Ethernet for Sensor and Display Applications: SerDes and Ethernet for Sensor and Display Applications*. Munich, Germany: Carl Hanser Verlag, 2023. ISBN: 978-3-446-46918-1. DOI: 10.3139/9783446470422.

[52] Abhineet Singh Tomar. "Modern Electrical/Electronic Infrastructure for Commercial Trucks: Generic Input/Output nodes for sensors and actuators in Commercial Trucks". Master's Thesis. Stockholm, Sweden: KTH Royal Institute of Technology, 2017.

[53] Hailong Zhu et al. "Requirements-Driven Automotive Electrical/Electronic Architecture: A Survey and Prospective Trends". In: *IEEE Access* 9 (2021), pp. 100096–100112. DOI: 10.1109/ACCESS.2021.3093077.

[54] Fedor Smirnov. "Design and Evaluation of Ethernet-based E/E-Architectures for Latency- and Safety-critical Applications: Entwurf und Evaluierung Ethernet-basierter E/E-Architekturen für latenz- und sicherheitskritische Anwendungen". Doctoral Dissertation. Nuremberg, Germany: Friedrich-Alexander-Universität Erlangen-Nürnberg.

[55] Hans-Joachim Faul, Simon Ramer and Markus Eckel. "Relays, contactors, cables, and connectors". In: *Lithium-ion batteries*. Ed. by Reiner Korthauer. Berlin and Heidelberg: Springer, 2018, pp. 133–154. ISBN: 978-3-662-53069-6. DOI: 10.1007/978-3-662-53071-9_12.

[56] Michael Trzesniowski. *Complete vehicle*. Wiesbaden: Springer Fachmedien Wiesbaden, 2023. ISBN: 978-3-658-39666-4. DOI: 10.1007/978-3-658-39667-1.

[57] Hans-Hermann Braess and Ulrich Seiffert, eds. *Vieweg Handbuch Kraftfahrzeugtechnik*. 7., aktualisierte Auflage. ATZ/MTZ-Fachbuch. Wiesbaden: Springer Vieweg, 2013. ISBN: 978-3-658-01690-6. DOI: 10.1007/978-3-658-01691-3.

[58] VDI/VDE. *Development of mechatronic and cyber-physical systems*. Berlin, 2021-11-01.

[59] Andreas Kreß, Markus Eberhardt and Jens Donig. “A Requirements Engineering Approach to Accelerate Innovation in Automotive Electronics”. In: *Automotive meets electronics*. Ed. by Michael Wahl. GMM-Fachbericht. Berlin and Offenbach: VDE Verlag GmbH, 2017, pp. 90–95. ISBN: 3800743698.

[60] Yutaka Onuma, Yoshiaki Terashima and Ryozo Kiyohara. “ECU Software Updating in Future Vehicle Networks”. In: *2017 31st International Conference on Advanced Information Networking and Applications Workshops (WAINA)*. IEEE, 2017, pp. 35–40. ISBN: 978-1-5090-6231-7. DOI: 10.1109/WAINA.2017.45.

[61] Oliver Creighton et al. *Towards Computer Aided, Iterative TSN and Ethernet based E/E Architecture Design*. Munich, Germany, September 15-16, 2020. URL: https://standards.ieee.org/wp-content/uploads/import/documents/other/eipatd-presentations/2020/D2-04-Creighton-Navet-Towards-Computer-Aided-Iterative-TSN-and-Ethernet-based-EE-updated.pdf.

[62] Nils Janßen et al. “Next Generation Automotive Zonal Architecture”. In: *AmE 2023 - Automotive Meets Electronics*. Ed. by Gesellschaft Mikroelektronik, Mikro- und Feinwerktechnik. Berlin, Offenbach and Piscataway, NJ: VDE Verlag GmbH and IEEE, 2023, pp. 17–21.

[63] Neha Nikesh Surjekar, Yutika Patwardhan and Vinaykumar Konduju. “A Case Study on Migrating Towards Funtionally Safe Zonal Architecture Using MBSE”. In: *INCOSE International Symposium* 33.1 (2023), pp. 1403–1417. ISSN: 2334-5837. DOI: 10.1002/iis2.13089.

[64] Josef Gebauer. “Ein modellbasiertes, graphisch notiertes, integriertes Verfahren zur Bewertung und zum Vergleich von Elektrik/Elektronik-Architekturen”. [Doctoral Dissertation]. Karlsruhe, Germany: Karlsruher Institut für Technologie, 2016.

[65] Péter Horváth. *Controlling*. 9. edition. 2003. ISBN: 978-3-8006-2992-5.

[66] Stevens S. S. “On the Theory of Scales of Measurement”. In: *Science* 103 (1946), pp. 677–680.

[67] Horst Zuse. *A Framework of Software Measurement*. Berlin, Germany: De Gruyter, 1998. ISBN: 978-3110155877.

[68] Patrick Suppes et al. *Foundations of Measurement: Volume 2: Geometrical, Threshold, and Probalistic Representation*. Academic Press, 2014. ISBN: 978-1493304424.

[69] Grazia Cecere, Nicoletta Corrocher and Marco Guerzoni. "Price or performance? A probabilistic choice analysis of the intention to buy electric vehicles in European countries". In: *Energy Policy* 118 (2018), pp. 19–32. ISSN: 03014215. DOI: 10.1016/j.enpol.2018.03.034.

[70] Lorenzo Nicoletti et al. "Design Parameters for the Early Development Phase of Battery Electric Vehicles". In: *World Electric Vehicle Journal* 11.3 (2020), p. 47. DOI: 10.3390/wevj11030047.

[71] Dan Scott and Ulrike Hoff. *Solving the EE dilemma of electric and autonomous vehicles*. Ed. by Siemens Digital Industries Software. 2019. URL: https://www.plm.automation.siemens.com/media/global/es/White%20Paper%20-%20Siemens%20SW%20Solving%20the%20EE%20dilemma%20of%20electric%20and%20autonomous%20vehicles%20WP-latest_tcm54-69007.pdf.

[72] James Folkestad and Russell Johnson. "Resolving the Conflict between Design and Manufacturing: Integrated Rapid Prototyping and Rapid Tooling (IRPRT)". In: *Journal of Industrial Technology* August 2001 to October 2001.Volume 17, Number 4 (2001).

[73] Stefan Raue. "Systemorientierung in der modellbasierten modularen E/E-Architekturentwicklung". Doctoral Dissertation. Eberhards Karls University Tübungen, 2018.

[74] Stefan Schumi and Alfons Graf. "Energy and Supply Concepts for Automated Driving". In: *Automotive meets electronics*. Ed. by Michael Wahl. GMM-Fachbericht. Berlin and Offenbach: VDE Verlag GmbH, 2018, pp. 95–99. ISBN: 3800745240.

[75] Mihir Mody et al. "Understanding Vehicle E/E Architecture Topologies for Automated Driving: System Partitioning and Tradeoff Parameters". In: *Electronic Imaging* 30.17 (2018), pp. 358-1-358–5. DOI: 10.2352/ISSN.2470-1173.2018.17.AVM-358.

[76] Michael Riesener et al. "Quantification of complexity in cyber-physical systems based on key figures". In: *Procedia CIRP* 100 (2021), pp. 445–450. ISSN: 22128271. DOI: 10.1016/j.procir.2021.05.102.

[77] Kaushik Sinha. "Structural Complexity and its Implications for Design of Cyber-Physical Systems". Doctoral Dissertation. Massachusetts, USA: Massachusetts Institute of Technology, 2014.

[78] Wei Wei. "Complexity management of vehicle wiring harnesses: An optimized model to analyse trade-offs between product and manufacturing costs". Master Thesis. Windsor, Ontario, Canada: University of Windsor, 2012.

[79] Lorenzo Nicoletti et al. "An Estimation of the Lightweight Potential of Battery Electric Vehicles". In: *Energies* 14.15 (2021), p. 4655. DOI: 10.3390/en14154655.

[80] Lorenzo Nicoletti. "Parametric Modeling of Battery Electric Vehicles in the Early Development Phase". Doctoral Dissertation. Munich, Germany: Technical University of Munich, 2022.

[81] Ganesh Sethuraman. "Vehicle Component Configuration Design and Packaging in Virtual Environment for Autonomous Electric Buses". Doctoral Dissertation. Tokyo, Japan: Tokyo University of Agriculture and Technology, 2020.

[82] Stephan Fuchs. *Verfahren zur parameterbasierten Gewichtsabschätzung neuer Fahrzeugkonzepte: Doctoral Thesis*. 2014. URL: https://mediatum.ub.tum.de/?id=1207264.

[83] Roeland M. M. Hogt and Frank G. Rieck. "Electric Vehicle Packaging Tool (EVPT), validation and application". In: *2013 World Electric Vehicle Symposium and Exhibition (EVS27)*. IEEE, 2013, pp. 1–12. ISBN: 978-1-4799-3832-2. DOI: 10.1109/EVS.2013.6914787.

[84] Kai Kuchenbuch. "Methodik zur Identifikation und zum Entwurf packageoptimierter Elektrofahrzeuge". Dissertation. Technische Universität Braunschweig, 2012.

[85] Johannes M. Stocker. *Standardisierung der Rahmenanbauteile-Topologie bei Nutzfahrzeugen: Doctoral Dissertation*. Munich: Verlag Dr. Hut, 2019. ISBN: 978-3-8439-4278-2.

[86] Armin Förg. *Architekturplanung und virtueller Fahrzeugentwurf schwerer Nutzfahrzeuge in der frühen Konzeptphase: Doctoral Dissertation*. Munich: Dr. Hut, 2020. ISBN: 978-3-8439-4605-6.

[87] Martin Bornemann. *Zone Controllers Build Bridge to Tomorrow's Technology: [White Paper]*. 2021. URL: `https://www.aptiv.com/docs/default-source/white-papers/2021_aptiv_whitepaper_zonecontroller.pdf?sfvrsn=c1f9163d_23` (visited on 18/04/2023).

[88] Infineon Technologies AG. *Dependable Solutions for Future Mobility and Zonal E/E Architectures*. URL: `https://www.infineon.com/dgdl/Infineon-Dependability_and_Zonal_Architectures-Presentations-v01_00-EN.pdf?fileId=8ac78c8c7d0d8da4017d27fe2a1634f6`.

[89] Jochen Klaus-Wagenbrenner. *Zonal EE Architecture: Towards a Fully Automotive Ethernet–Based Vehicle Infrastructure*. 2019-09-24. URL: `https://standards.ieee.org/wp-content/uploads/import/documents/other/eipatd-presentations/2019/D1-04_KLAUS-Zonal_EE_Architecture.pdf`.

[90] Kirill Gorelik, Ahmet Kilic and Roman Obermaisser. "Connected Energy Management System for Automated Electric Vehicles With Fail-Operational Powertrain and Powernet". In: *IEEE Transactions on Vehicular Technology* 68.10 (2019), pp. 9588–9603. ISSN: 0018-9545. DOI: `10.1109/TVT.2019.2934777`.

[91] Peter Aberl and et al. *How a Zone Architecture Paves the Way to a Fully Software-Defined Vehicle: [White Paper]*. 2023. URL: `https://www.ti.com/lit/wp/spry345a/spry345a.pdf?ts=1681378174830&ref_url=https%253A%252F%252Fwww.google.com%252F` (visited on 18/04/2023).

[92] Greg Avery and et al. "ZONAL ARCHITECTURE: MAKING THE CAR OF THE FUTURE POSSIBLE: [White paper]". In: 2023 (). URL: `https://engage.molex.com/Zonal-Architecture-Whitepaper.html`.

[93] Gereon Meyer. "Schlüsselrolle der EE-Architektur und der Bordnetze für das Automobil der Zukunft: [White Paper]". In: (2020). URL: https://www.strategiekreis-automobile-zukunft.de/files/enova-position-bordnetze-und-fahrzeugarchitektur.pdf.

[94] Michael Niklas-Höret. *Impact of the Zone Architecture on the in Vehicle SW Distribution*. URL: https://ahkfinnland.de/filehub/deliverFile/e9a3077f-b2ce-425f-8835-a61b2e62513b/1229026/Continental_Keynote_1229026.pdf (visited on 18/04/2023).

[95] Maurice Müller. "Zonen-basierte E/E-Architektur im Fokus: AUTOMATISIERTES UND VERNETZTES FAHREN". In: *HANSER automotive* 4 (2021), pp. 26–28. URL: https://www.hanser-automotive.de/a/fachartikel/zonen-basierte-ee-architektur-im-fokus-342944.

[96] Mario Maul, Gerhard Becker and Ulrich Bernhard. "Service-oriented EE zone architecture key elements for new market segments". In: *ATZelektronik worldwide* 13.1 (2018), pp. 36–41. DOI: 10.1007/s38314-017-0092-4.

[97] Michael Kearns, Yishay Mansour and Andrew Y. Ng. "An Information-Theoretic Analysis of Hard and Soft Assignment Methods for Clustering". In: *Learning in Graphical Models*. Ed. by Michael I. Jordan. Nato Science Series D. Dordrecht: Springer, 1998, pp. 495–520. ISBN: 978-94-010-6104-9. DOI: 10.1007/978-94-011-5014-9_18.

[98] Mohammed J. Zaki and Wagner Meira. *Data mining and analysis: Fundamental concepts and algorithms*. New York, NY: Cambridge Univ. Press, 2014. ISBN: 9780521766333.

[99] Frank Nielsen. "Hierarchical Clustering". In: *Introduction to HPC with MPI for data science*. Ed. by Frank Nielsen. Undergraduate Topics in Computer Science. Cham and Heidelberg: Springer, 2016, pp. 195–211. ISBN: 978-3-319-21902-8. DOI: 10.1007/978-3-319-21903-5_8.

[100] Martin. Ester et al. "A density-based algorithm for discovering clusters in large spatial databases with noise". In: *International Conference on Knowledge Discovery & Data Mining*. Ed. by Evangelos Simoudis. Menlo Park, Calif.: AAAI Press, 1996. ISBN: 1577350049.

[101] Mihael Ankerst et al. "OPTICS". In: *Proceedings of the 1999 ACM SIGMOD international conference on Management of data*. Ed. by Susan B. Davidson. ACM Conferences. New York, NY: ACM, 1999, pp. 49–60. ISBN: 1581130848. DOI: 10.1145/304182.304187.

[102] Xiaowei Xu et al. "A distribution-based clustering algorithm for mining in large spatial databases". In: *Proceedings / 14th International Conference on Data Engineering, February 23-27, 1998, Orlando, Florida*. Los Alamitos, Calif.: IEEE Computer Soc. Pr, 1998, pp. 324–331. ISBN: 0-8186-8289-2. DOI: 10.1109/ICDE.1998.655795.

[103] Xavier Dominguez, Paola Mantilla-Perez and Pablo Arboleya. "Toward Smart Vehicular dc Networks in the Automotive Industry: Process, computational tools, and trends in the design and simulation of vehicle electrical distribution systems". In: *IEEE Electrification Magazine* 8.1 (2020), pp. 61–68. ISSN: 2325-5897. DOI: 10.1109/MELE.2019.2962890.

[104] Huong Giang Nguyen, Marlene Kuhn and Jörg Franke. "Manufacturing automation for automotive wiring harnesses". In: *Procedia CIRP* 97 (2021), pp. 379–384. ISSN: 22128271. DOI: 10.1016/j.procir.2020.05.254.

[105] Ruddy Guerrero et al. "Simulation Model for Wire Harness Design in the Car Production Line Optimization Using the SimPy Library". In: *Sustainability* 14.12 (2022), p. 7212. DOI: 10.3390/su14127212.

[106] Steve Carlson. *Reduced Twisted Pair Gigabit Ethernet PHY Call for Interest*. 2012-03-12.

[107] Simon Fürst et al. "Digitalisierung/ Elektrik/ Elektronik/ Software". In: *Vieweg Handbuch Kraftfahrzeugtechnik*. Ed. by Stefan Pischinger and Ulrich Seiffert. Wiesbaden: Springer Fachmedien Wiesbaden, 2021, pp. 861–1008. ISBN: 978-3-658-25556-5. DOI: 10.1007/978-3-658-25557-2_7.

[108] Elisabeth Oberleithner. "Applicability of Automotive Powerline Communication A Theory- and Practice-based Study". Doctoral Thesis. Munich, Germany: Technical University of Munich, 2019-04-01. URL: https://mediatum.ub.tum.de/doc/1481898/1481898.pdf.

[109] Fujikura Ltd. *Automotive: Fujikura offers a variety of electricity distribution systems for vehicles*. Ed. by Fujikura Ltd. 2023. URL: https://www.fujikura.com/solutions/automotive/ (visited on 26/07/2023).

[110] Stefan Pischinger and Ulrich Seiffert, eds. *Vieweg Handbuch Kraftfahrzeugtechnik*. Wiesbaden: Springer Fachmedien Wiesbaden, 2021. ISBN: 978-3-658-25556-5. DOI: 10.1007/978-3-658-25557-2.

[111] Rajeeth Tharma, Roland Winter and Martin Eigner. "AN APPROACH FOR THE IMPLEMENTATION OF THE DIGITAL TWIN IN THE AUTOMOTIVE WIRING HARNESS FIELD". In: *Proceedings of the DESIGN 2018 15th International Design Conference*. Design Conference Proceedings. Faculty of Mechanical Engineering and Naval Architecture, University of Zagreb, Croatia and The Design Society, Glasgow, UK, 2018, pp. 3023–3032. DOI: 10.21278/idc.2018.0188.

[112] Kai Borgeest. *Elektronik in der Fahrzeugtechnik: Hardware, Software, Systeme und Projektmanagement*. 4., aktualisierte und erweiterte Auflage. ATZ/MTZ-Fachbuch. Wiesbaden, Germany: Springer Vieweg, 2021. ISBN: 978-3-658-23663-2. DOI: 10.1007/978-3-658-23664-9.

[113] Maja Diebig. "Entwicklung einer Methodik zur simulationsbasierten Dimensionierung von Kfz-Bordnetzen". Doctoral Thesis. Technische Universität Dortmund, 2016.

[114] SPE Industrial Partner Network e.V. *Single Pair Ethernet: Applications and Use Cases Studies*. 2023. URL: https://www.single-pair-ethernet.com/en/artikel/single-pair-ethernet-use-cases.

[115] Hisup Park, Soo-Hong Lee and Mark R. Cutkosky. "Computational Support for Concurrent Engineering of Cable Harnesses". In: DOI: 10.1115/CIE1992-0032.

[116] Chung-Wei Lin et al. "Efficient Wire Routing and Wire Sizing for Weight Minimization of Automotive Systems". In: *IEEE Transactions on Computer-Aided Design of Integrated Circuits and Systems* 34.11 (2015), pp. 1730–1741. ISSN: 0278-0070. DOI: 10.1109/TCAD.2015.2448680.

[117] Ta-Yang Huang et al. “Intra-vehicle network routing algorithm for wiring weight and wireless transmit power minimization”. In: *The 20th Asia and South Pacific Design Automation Conference*. IEEE, 2015, pp. 273–278. ISBN: 978-1-4799-7792-5. DOI: 10.1109/ASPDAC.2015.7059017.

[118] Satyan Chandra et al. “WIRING SYSTEM ARCHITECTURE”. US 2019 / 0217794 A1. 2019. URL: https://patentimages.storage.googleapis.com/7f/ea/45/f221d47693fe6e/US20190217794A1.pdf.

[119] Jerome Trommnau et al. “Overview of the State of the Art in the Production Process of Automotive Wire Harnesses, Current Research and Future Trends”. In: *Procedia CIRP* 81 (2019), pp. 387–392. ISSN: 22128271. DOI: 10.1016/j.procir.2019.03.067.

[120] Nafiseh Masoudi. “Geometric-based Optimization Algorithms for Cable Routing and Branching in Cluttered Environments”. Doctoral Dissertation. Clemson: Clemson University, South Carolina, 2020-08-01. URL: https://tigerprints.clemson.edu/all_dissertations/2702/.

[121] Sjon Moore. “Modeling and Optimizing Wire Harness Costs for Variation Complexity”. In: *SAE Technical Paper Series*. SAE Technical Paper Series. SAE International400 Commonwealth Drive, Warrendale, PA, United States, 2016. DOI: 10.4271/2016-01-0107.

[122] Zaoxu Zhu. “Automatic 3D Routing for the Physical Design of Electrical Wiring Interconnection Systems for Aircraft”. Doctoral Dissertation. Delft: Delft University of Technology, 2016. DOI: 10.4233/uuid:2ca107b4-202d-4638-a044-d45649b89275.

[123] Tobias Karlsson. “Optimization of Cable Harness Routing: Mathematical Modelling, Lagrangian Relaxation, and Subgradient Optimization”. Master thesis. Gothenburg, Sweden: Chalmers University of Technology, 2020. URL: https://hdl.handle.net/20.500.12380/301802.

[124] Saekyeol Kim et al. "Automatic design system for generating routing layout of tubes, hoses, and cable harnesses in a commercial truck". In: *Journal of Computational Design and Engineering* 8.4 (2021), pp. 1098–1114. DOI: 10.1093/jcde/qwab034.

[125] Marco Mandolini et al. *A Design to Cost Method for Electric Cable Harness*. Vancouver, Canada, 21-25 August, 2017. URL: https://www.researchgate.net/publication/319504238_A_design_to_cost_method_for_electric_cable_harness.

[126] Yingfeng Zhao et al. "Multi-Branch Cable Harness Layout Design Based on Genetic Algorithm with Probabilistic Roadmap Method". In: *Chin. J. Mech. Eng.* 34.33 (2021). DOI: 10.1186/s10033-021-00544-9.

[127] Konrad Zuse. *Der Plankalkül*. St. Augustin, Germany, 1972.

[128] E. W. Dijkstra. "A note on two problems in connexion with graphs". In: *Numer. Math.* 1 (1959), pp. 269–271. DOI: 10.1007/BF01386390.

[129] Peter Hart, Nils Nilsson and Bertram Raphael. "A Formal Basis for the Heuristic Determination of Minimum Cost Paths". In: *IEEE Transactions on Systems Science and Cybernetics* 4.2 (1968), pp. 100–107. ISSN: 0536-1567. DOI: 10.1109/TSSC.1968.300136.

[130] Uwe Koehler. "Lithium-ion battery system design". In: *Lithium-ion batteries*. Ed. by Reiner Korthauer. Berlin and Heidelberg: Springer, 2018, pp. 89–100. ISBN: 978-3-662-53069-6. DOI: 10.1007/978-3-662-53071-9_8.

[131] B. Scrosati, J. Garche and W. Tillmetz. *Advances in Battery Technologies for Electric Vehicles*. Number 80. Woodhead Publishing Series in Energy. Cambridge, United Kingdom: Elsevier Ltd., 2015. ISBN: 978-1-78242-377-5.

[132] Leif E. Asp and et al. "A Structural Battery and its Multifunctional Performance". In: *Advanced Energy and Sustainability Research* 2.3 (2021), p. 2000093. ISSN: 2699-9412. DOI: 10.1002/aesr.202000093.

[133] M.F.R. Zwicker et al. "Automotive battery pack manufacturing – a review of battery to tab joining". In: *Journal of Advanced Joining Processes* 1 (2020), p. 100017. ISSN: 26663309. DOI: 10.1016/j.jajp.2020.100017.

[134] Steffen Link, Christoph Neef and Tim Wicke. "Trends in Automotive Battery Cell Design: A Statistical Analysis of Empirical Data". In: *Batteries* 9.5 (2023), p. 261. DOI: 10.3390/batteries9050261.

[135] Michael Kries et al. *An Overview of Costs for Vehicle Components, Fuels, Greenhouse Gas Emissions and Total Cost of Ownership Update 2017: Research Report - UCD-ITS-RR-17-45*. Ed. by Institute of Transportation Studies, University of California. California, USA, 2017.

[136] Hendrik Löbberding et al. "From Cell to Battery System in BEVs: Analysis of System Packing Efficiency and Cell Types". In: *World Electric Vehicle Journal* 11.4 (2020), p. 77. DOI: 10.3390/wevj11040077.

[137] Kaveh Sarrafan et al. "Accurate range estimation for an electric vehicle including changing environmental conditions and traction system efficiency". In: *IET Electrical Systems in Transportation* 7.2 (2017), pp. 117–124. ISSN: 2042-9738. DOI: 10.1049/iet-est.2015.0052.

[138] Ali Dinc and Murat Otkur. *ICMAE 2020: 2020 11th International Conference on Mechanical and Aerospace Engineering : July 14-17, 2020, Athens, Greece*. Piscataway, NJ: IEEE, 2020. ISBN: 9781728183220. DOI: 10.1109/ICMAE50897.2020. URL: http://ieeexplore.ieee.org/servlet/opac?punumber=9171913.

[139] Nansi Xue. "Design and Optimization of Lithium-Ion Batteries for Electric-Vehicle Applications". Doctoral Dissertation. Michigan, USA: University of Michigan, 2014. URL: https://hdl.handle.net/2027.42/107299.

[140] Matthias Manuel Kerler. *Eine Methode zur Bestimmung der optimalen Zellgröße für Elektrofahrzeuge: Doctoral Dissertation*. München: Dr. Hut, 2019. ISBN: 978-3-8439-3948-5.

[141] Felix Goudreault et al. “LiDAR-in-the-Loop Hyperparameter Optimization”. In: *2023 IEEE/CVF Conference on Computer Vision and Pattern Recognition (CVPR)*. Ed. by IEEE. 2023, pp. 13404–13414. ISBN: 979-8-3503-0129-8.

[142] Oliver Kramer. *Genetic Algorithm Essentials*. Vol. 679. Cham: Springer International Publishing, 2017. ISBN: 978-3-319-52155-8. DOI: 10.1007/978-3-319-52156-5.

[143] Iman Rahimi et al. “A comparative study on evolutionary multi-objective algorithms for next release problem”. In: *Applied Soft Computing* 144 (2023), p. 110472. ISSN: 15684946. DOI: 10.1016/j.asoc.2023.110472.

[144] Heide Budde-Meiwes et al. “A review of current automotive battery technology and future prospects”. In: *Proceedings of the Institution of Mechanical Engineers, Part D: Journal of Automobile Engineering* 227.5 (2013), pp. 761–776. ISSN: 0954-4070. DOI: 10.1177/0954407013485567.

[145] Peter Kurzweil and Otto Dietlmeier. *Elektrochemische Speicher: Superkondensatoren, Batterien, Elektrolyse-Wasserstoff, rechtliche Rahmenbedingungen*. 2., aktualisierte und erweiterte Auflage. Lehrbuch. Wiesbaden and Heidelberg: Springer Vieweg, 2018. ISBN: 3658218282. DOI: 10.1007/978-3-658-21829-4.

[146] H. Helmholtz. “Studien über electrische Grenzschichten”. In: *Annalen der Physik und Chemie* 243.7 (1879), pp. 337–382. ISSN: 00033804. DOI: 10.1002/andp.18792430702.

[147] Yu-En Wu and Chen-Han Tai. “Novel Bidirectional Isolated DC/DC Converter with High Gain Ratio and Wide Input Voltage for Electric Vehicle Storage Systems”. In: *Batteries* 8.11 (2022), p. 240. DOI: 10.3390/batteries8110240.

[148] Maobing Li et al. “The structure and control method of hybrid power source for electric vehicle”. In: *Energy* 112 (2016), pp. 1273–1285. ISSN: 03605442. DOI: 10.1016/j.energy.2016.06.009.

[149] Lisa Braun et al. "Using a Domain Model to Precisely Describe function-agnostic Electric/Electronic-Architectures". In: *Automotive meets electronics*. Ed. by Michael Wahl. GMM-Fachbericht. Berlin and Offenbach: VDE Verlag GmbH, 2017, pp. 59–64. ISBN: 3800743698.

[150] Philipp Obergfell et al. "Viewpoint-Based Methodology for Adaption of Automotive E/E-Architectures". In: *2018 IEEE International Conference on Software Architecture Companion (ICSA-C)*. IEEE, 2018, pp. 128–135. ISBN: 978-1-5386-6585-5. DOI: 10.1109/ICSA-C.2018.00041.

[151] Jonas Maier and Hans-Christian Reuss. "Design of Zonal E/E Architectures in Vehicles Using a Coupled Approach of k-Means Clustering and Dijkstra's Algorithm". In: *Energies* 16 (2023), p. 6884. DOI: 10.3390/en16196884.

[152] Ricardo Gonzalez de Oliveira, Nicolas Navet and Achim Henkel. "Multi-Objective Optimization for Safety-Related Available E/E Architectures Scoping Highly Automated Driving Vehicles". In: *ACM Transactions on Design Automation of Electronic Systems* 28.3 (2023), pp. 1–37. ISSN: 1084-4309. DOI: 10.1145/3582004.

[153] Weiss Gereon, Schleiss Philipp and Drabek Christian. *Towards Flexible and Dependable E/E-Architectures for Future Vehicles*. Göteborg, Sweden, 2016-09-01.

[154] Matthias Traub, Alexander Maier and Kai L. Barbehon. "Future Automotive Architecture and the Impact of IT Trends". In: *IEEE Software* 34.3 (2017), pp. 27–32. ISSN: 0740-7459. DOI: 10.1109/MS.2017.69.

[155] Alessio Bucaioni and Patrizio Pelliccione. "Technical Architectures for Automotive Systems". In: *2020 IEEE International Conference on Software Architecture (ICSA)*. IEEE, 2020, pp. 46–57. DOI: 10.1109/ICSA47634.2020.00013.

[156] Thomas Liebetrau. "E/E Architecture Transformation How it impacts value chain and networking technologies: Automotive meets electronics : 13. GMM-Symposium, September 29-30, 2022 in Dortmund". In: *AmE 2022*. Ed. by Michael Wahl and Anne Stockem. Vol. 104. GMM-Fachbericht. Berlin and Offenbach: VDE Verlag GmbH, 2022, pp. 33–39. ISBN: 9783800759552.

[157] Bernhard Rill and James Scobie. "Real-time Software Enablement in Zonal Controllers". In: *ATZelectronics worldwide* 18 (2023), pp. 8–13. DOI: 10.1007/s38314-023-1478-0.

[158] Dr. Ing. h.c. F. Porsche AG. *Porsche fine-tunes E-Performance for future series models: PPE Insights: Premium platform electric in detail.* 2023. URL: https://media.porsche.com/mediakit/battery-electric-vehicle-workshop/en/bev-workshop/insight-ppe-new-platform-for-upcoming-series-models.

[159] Paul S. Bradley and Usama M. Fayyad. "Refining Initial Points for K-Means Clustering". In: *Proceedings of the 15th International Conference on Machine Learning (ICML98)*. Ed. by Morgan Kaufmann Publishers Inc. 1998, pp. 91–99. ISBN: 978-1-55860-556-5.

[160] Alan R.R. de Freitas, Peter J. Fleming and Frederico G. Guimarães. "Aggregation Trees for visualization and dimension reduction in many-objective optimization". In: *Information Sciences* 298 (2015), pp. 288–314. ISSN: 00200255. DOI: 10.1016/j.ins.2014.11.044.

[161] Shanu Verma, Millie Pant and Vaclav Snasel. "A Comprehensive Review on NSGA-II for Multi-Objective Combinatorial Optimization Problems". In: *IEEE Access* 9 (2021), pp. 57757–57791. DOI: 10.1109/ACCESS.2021.3070634.

[162] V. Khare, X. Yao and K. Deb. "Performance Scaling of Multi-objective Evolutionary Algorithms". In: *Evolutionary Multi-Criterion Optimization*. Ed. by Gerhard Goos. Vol. 2632. Lecture Notes in Computer Science. Berlin/Heidelberg: Springer Berlin Heidelberg, 1999, pp. 376–390. ISBN: 978-3-540-01869-8. DOI: 10.1007/3-540-36970-8_27.

[163] K. Deb et al. "A fast and elitist multiobjective genetic algorithm: NSGA-II". In: *IEEE Transactions on Evolutionary Computation* 6.2 (2002), pp. 182–197. ISSN: 1089778X. DOI: 10.1109/4235.996017.

[164] Dimas de La Fuente, Miguel A. Vega-Rodríguez and Carlos J. Pérez. "Automatic selection of a single solution from the Pareto front to identify key players in social networks". In: *Knowledge-Based Systems* 160 (2018), pp. 228–236. ISSN: 09507051. DOI: 10.1016/j.knosys.2018.07.018.

[165] S. Arora and A. Kapoor. "Mechanical Design and Packaging of Battery Packs for Electric Vehicles". In: *Behaviour of Lithium-Ion Batteries in Electric Vehicles*. Ed. by Gianfranco Pistoia and Boryann Liaw. Green Energy and Technology. Cham: Springer International Publishing, 2018, pp. 175–200. ISBN: 978-3-319-69949-3. DOI: 10.1007/978-3-319-69950-9_8.

[166] Lorenzo Nicoletti et al. "Topology analysis of electric vehicles, with a focus on the traction battery". In: *Forschung im Ingenieurwesen* 85.2 (2021), pp. 457–467. ISSN: 0015-7899. DOI: 10.1007/s10010-020-00422-1.

[167] Marco Münster. *UMV: NEXT GENERATION CAR: Urban Modular Vehicle*. Ed. by German Aerospace Center, Institute of Vehicle Concepts. 2019.

[168] Alexander Böhm and Manfred Kraus. "The Rolling Chassis from Schaeffler Spearheads New Mobility Solutions". In: *Schaeffler Kolloqium 2022: Digital conference book 2022*. Ed. by Schaeffler AG. 2022. URL: https://www.schaeffler.com/en/media/dates-events/kolloquium/digital-conference-book-2022/rolling-chassis/.

[169] Michael Lahres et al. *FlexCAR – eine autonome Fahrzeugforschungsplattform der Zukunft*. Stuttgart, Germany, 15-16. February 2023.

[170] Timo Woopen et al. "UNICARagil - Where We Are and Where We Are Going". In: *29. Aachen Colloquium Sustainable Mobility*. 2020, pp. 285–308. ISBN: 978-3-00-064871-7. DOI: 10.18154/RWTH-2020-10407.

[171] Kyung-Min Lee et al. *IECON 2019 - 45th Annual Conference of the IEEE Industrial Electronics Society: Convention Center, Lisbon, Portugal, 14-17 October, 2019 : proceedings*. Piscataway, NJ: IEEE, 2019. ISBN: 9781728148786. URL: http://ieeexplore.ieee.org/servlet/opac?punumber=8897531.

[172] Laurenz Tippe et al. "Modular Simulation of Zonal Architectures and Ring Topologies for Automotive Power Nets". In: *2021 IEEE Vehicle Power and Propulsion Conference (VPPC)*. IEEE, 2021, pp. 1–5. ISBN: 978-1-6654-0528-7. DOI: 10.1109/VPPC53923.2021.9699179.

[173] Lidiya Komsiyska et al. “Critical Review of Intelligent Battery Systems: Challenges, Implementation, and Potential for Electric Vehicles”. In: *Energies* 14.18 (2021), p. 5989. DOI: 10.3390/en14185989.

[174] Jonas Neckenich. “3D-Master-Leitungssatz - Konzept zur Entwicklung von Leitungssätzen als 3D-Master in einem realistischen, vollständigen DMU-Modell”. Doctoral Dissertation. Saarbrücken: Saarland University, 2017. DOI: 10.22028/D291-26986.

[175] Jonas Maier and Hans-Christian Reuss. *A Framework for Optimizing the Wiring Harness of Automated Electric Vehicles*. Shanghai, 1-2 December, 2022.

[176] R. L. Graham and Pavol Hell. “On the History of the Minimum Spanning Tree Problem”. In: *IEEE Annals of the History of Computing* 7.1 (1985), pp. 43–57. ISSN: 1058-6180. DOI: 10.1109/MAHC.1985.10011.

[177] William Siegle. “Optimierung von Kabelsträngen für zonale E/E-Architekturen in elektrischen Fahrzeugen: Optimization of Wiring Harnesses for Zonal E/E Architectures in Electric Vehicles”. Research Thesis. Stuttgart, Germany: University of Stuttgart, 2024.

[178] Lucas Mauser, Stefan Wagner and Peter Ziegler. “Methodical Approach for Centralization Evaluation of Modern Automotive E/E Architectures”. In: *Software Architecture. ECSA 2022 Tracks and Workshops*. Ed. by Thais Batista et al. Cham: Springer International Publishing, 2023, pp. 165–179. ISBN: 978-3-031-36888-2.

[179] Calin Iclodean, Nicolae Cordos and Bogdan Ovidiu Varga. “Autonomous Shuttle Bus for Public Transportation: A Review”. In: *Energies* 13.11 (2020), p. 2917. DOI: 10.3390/en13112917.

[180] Jonatan J. Gómez Vilchez, Roberto Pasqualino and Yeray Hernandez. “The new electric SUV market under battery supply constraints: Might they increase CO2 emissions?” In: *Journal of Cleaner Production* 383 (2023), p. 135294. ISSN: 09596526. DOI: 10.1016/j.jclepro.2022.135294.

[181] Shiqi Ou et al. “Relationships between Vehicle Pricing and Features: Data Driven Analysis of the Chinese Vehicle Market”. In: *Energies* 13.12 (2020), p. 3088. DOI: 10.3390/en13123088.

[182] Siddhant Gudhe and Sanjeev Singh. "Co-ordinated Control of Two Batteries for Traction Motor and Auxiliary Loads in Electric Vehicle". In: *2022 IEEE 2nd International Conference on Sustainable Energy and Future Electric Transportation (SeFeT)*. IEEE, 2022, pp. 1–6. ISBN: 978-1-6654-8057-4. DOI: 10.1109/SeFeT55524.2022.9908702.

[183] Yi-Hsiang Tseng and Yee-Pien Yang. "Torque and Battery Distribution Strategy for Saving Energy of an Electric Vehicle with Three Traction Motors". In: *Applied Sciences* 10.8 (2020), p. 2653. DOI: 10.3390/app10082653.

[184] Houssem Guissouma et al. "An Empirical Study on the Current and Future Challenges of Automotive Software Release and Configuration Management". In: *2018 44th Euromicro Conference on Software Engineering and Advanced Applications (SEAA)*. IEEE, 2018, pp. 298–305. ISBN: 978-1-5386-7383-6. DOI: 10.1109/SEAA.2018.00056.

[185] Subramaniam Jayanti, Yagnanarayanan Kalyanaraman and Karthik Ramani. "Shape-based clustering for 3D CAD objects: A comparative study of effectiveness". In: *Computer-Aided Design* 41.12 (2009), pp. 999–1007. ISSN: 00104485. DOI: 10.1016/j.cad.2009.07.003.

[186] Xujian Cui et al. "An application of evolutionary computation algorithm in multidisciplinary design optimization of battery packs for electric vehicle". In: *Energy Storage* 2.3 (2020). ISSN: 2578-4862. DOI: 10.1002/est2.158.

[187] Thomas Hogenmüller and Amir Bar-Niv. *Enhancing Automotive Ethernet Efficiency for Emerging, Asymmetrical Use Cases*. Japan, November 9-10, 2022. URL: https://standards.ieee.org/wp-content/uploads/2022/12/D2_09_Ami-Bar-Niv-Thomas-Hogenmuller_-Enhancing-Automotive-Ethernet-Efficiency-for-Emerging-Asymmetrical-Use-Cases.pdf.

[188] Steven D. Eppinger and Tyson R. Browning. *Design Structure Matrix Methods and Applications*. The MIT Press, 2012. DOI: 10.7551/mitpress/8896.001.0001.

[189] Hadi Askaripoor, Thilo Mueller and Alois Knoll. *E/E Designer: a Framework to Design and Synthesize Vehicle E/E Architecture*. 2023. DOI: 10.36227/techrxiv.23530803.v1.

[190] D. Buttle and S. Gold. "MCUs and Virtualization in Zone E/E Architectures". In: *ATZelectronics worldwide* 17 (2022), pp. 18–24. DOI: 10.1007/s38314-022-0819-8.

[191] Hyuksoo Jang et al. "Design of a Hybrid In-Vehicle Network Architecture Combining Zonal and Domain Architectures for Future Vehicles". In: *2023 IEEE 6th International Conference on Knowledge Innovation and Invention (ICKII)*. IEEE, 2023, pp. 33–37. ISBN: 979-8-3503-2353-5. DOI: 10.1109/ICKII58656.2023.10332574.

[192] Chengyu Cui, Chulsun Park and Sungkwon Park. "Physical Length and Weight Reduction of Humanoid In-Robot Network with Zonal Architecture". In: *Sensors (Basel, Switzerland)* 23.5 (2023). DOI: 10.3390/s23052627.

[193] Giorgio Sulligoi et al. "Design of Zonal Electrical Distribution Systems for Ships and Oil Platforms: Control Systems and Protections". In: *IEEE Transactions on Industry Applications* 56.5 (2020), pp. 5656–5669. ISSN: 0093-9994. DOI: 10.1109/TIA.2020.2999035.

[194] Richard Alexander, Danielle Meyer and Jiankang Wang. "A Comparison of Electric Vehicle Power Systems to Predict Architectures, Voltage Levels, Power Requirements, and Load Characteristics of the Future All-Electric Aircraft". In: *2018 IEEE Transportation Electrification Conference and Expo (ITEC)*. IEEE, 2018, pp. 194–200. ISBN: 978-1-5386-3048-8. DOI: 10.1109/ITEC.2018.8450240.

[195] James H. Gawron et al. "Life Cycle Assessment of Connected and Automated Vehicles: Sensing and Computing Subsystem and Vehicle Level Effects". In: *Environmental science & technology* 52.5 (2018), pp. 3249–3256. DOI: 10.1021/acs.est.7b04576.

[196] Matti Kutila et al. "Sensor System Power Adaptation for Automated Vehicles". In: *2019 IEEE 15th International Conference on Intelligent Computer Communication and Processing (ICCP)*. Ed. by Sergiu Nedevschi, Rodica Potolea and Radu Razvan Slavescu. Piscataway, NJ: IEEE, 2019. ISBN: 9781728149141.

[197] Tom P. Kohler. *Prädiktives Leistungsmanagement in Fahrzeugbordnetzen*. Wiesbaden, Germany: Springer Fachmedien Wiesbaden, 2014. ISBN: 978-3-658-05011-5. DOI: 10.1007/978-3-658-05012-2.

[198] Ilya Kavalchuk et al. "Advanced Simulation of Power Consumption of Electric Vehicles". In: *International Journal of Electrical, Computer, Electronics and Communication Engineering* 9.1 (2015), pp. 53–59.

[199] Henning Wallentowitz and Konrad Reif. *Handbuch Kraftfahrzeugelektronik: Grundlagen - Komponenten - Systeme - Anwendungen.* ATZ/MTZ-Fachbuch. Wiesbaden, Germany: Vieweg + Teubner Verlag, 2011. ISBN: 978-3-8348-0700-7.

[200] Paul Hochrein. *Leistungsoptimale Regelung von Hochstromverbrauchern im Fahrwerk: Zugl.: Kassel, Univ., Diss., 2013.* Kassel, Germany: Kassel University Press, 2013. ISBN: 9783862196043.

[201] Michael Radomir Fabis. "Beitrag zum Energiemanagement in Kfz-Bordnetzen". Doctoral Dissertation. Berlin, Germany: Technical University of Berlin, 2006.

[202] Robert Bosch GmbH. *Bosch Automotive Electrics and Automotive Electronics: Systems and Components, Networking and Hybrid Drive.* 5. Edition. Bosch Professional Automotive. Wiesbaden: Springer Vieweg, 2014. ISBN: 978-3658017835.

[203] Detlef Heinemann. "Strukturen von Batterie- und Energiemanagementsystemen mit Bleibatterien und Ultracaps". Doctoral Dissertation. Berlin, Germany: Technical University of Berlin, 2006.

[204] SAB Bröckskes GmbH & Co. KG. *CATLine SPE Robot: Single-Pair-Ethernet cable, suitable for robots with UL recognition: [Catalog].* 2024. URL: https://www.sab-kabel.de/produkte/buskabel-und-leitungen/catline-spe-robot.html (visited on 03/01/2024).

[205] Lapp Tannehill Inc. *Automotive Ethernet Cables: [Catalog].* 2024. URL: https://www.lapptannehill.com/wire-cable/automotive-wire-cable/automotive-ethernet-cable (visited on 03/01/2024).

[206] Kusum Deep et al. "A real coded genetic algorithm for solving integer and mixed integer optimization problems". In: *Applied Mathematics and Computation* 212.2 (2009), pp. 505–518. ISSN: 00963003. DOI: 10.1016/j.amc.2009.02.044.

[207] The MathWorks, Inc. *Genetic Algorithm Options*. 2023. URL: https://de.mathworks.com/help/gads/genetic-algorithm-options.html#f6633.

[208] Hans-Martin Fischer. *Spannungsklassen in der Elektromobilität, Ausgabe 2021*. Ed. by ZVEI - German Electrical and Electronic Manufacturers' Association. 2021. URL: https://www.zvei.org/fileadmin/user_upload/Presse_und_Medien/Publikationen/2021/Juli/ZVEI-Spannungsklassen-in-der-Elektromobilitaet_Ausgabe-2021/ZVEI-Spannungsklassen-in-der-Elektromobilitaet_Ausgabe-2021.pdf.

Glossary

Adjacency matrix	The adjacency matrix represents a graph respectively the connections or edges between the vertices. Therefore it is a symmetrical square matrix with entries either one for a connection between two vertices or zero if there is no connection.
Bus/Data-Bus	A system, for transferring data and/or power between hardware and software components.
ISO26262	An International Organization for Standardization (ISO) standard that includes methods and products for the development of safety-related electrical/electronic systems. The standard has been valid since 2011.
Powertrain	The powertrain comprises all components converting the energy, both electrical and mechanical, from the energy storage system and transmit it to the wheels of the vehicle.
SysML	SysML is a modeling language mainly used in systems engineering. It defines objects and interfaces of complex systems by using a graphic notation that is based on the UML 2.
UNICARagil	UNICARagil is a Federal Ministry of Education and Research founded project for developing new hard- and software components, plattforms and architectures for automated electric vehicles. Further information about the project can be found at www.unicaragil.de/en.

J. Maier, *Optimization of Zonal Electric/Electronic Architectures Using a Coupled Modeling Approach of Package and Wiring Harness*, Wissenschaftliche Reihe Fahrzeugtechnik Universität Stuttgart, https://doi.org/10.1007/978-3-658-48188-9

VDI2206	VDI2206 is a standard by the Association of German Engineers defining the development of mechatronic and cyber-physical systems using the V-model.
WLTP	The Worldwide harmonized Light vehicles Test Procedure is a global standard test cycle for consistent determination of efficiency and pollutants of passenger cars. The standard is established in the European Union, Iceland, Israel , Liechtenstein, Norway. Switzerland and Turkey.

Appendix

A. Vehicle Data

A.1 Battery Mass and Volume in Relation to Vehicle Data

Table A.1: Traction battery mass and volume in relation to vehicle data

Vehicle type	E_{bat} in Ah	m_{bat} in kg	m_{veh} in kg	$\frac{m_{bat}}{m_{veh}}$ in %	V_{bat} in m^3	V_{crg} in m^3	$\frac{V_{bat}}{V_{crg}}$ in %
BMW i3	120.0	278	1345	0.207	0.259	0.260	0.996
Nissan Leaf	62.0	410	1580	0.259	0.483	0.435	1.111
Tesla Model 3	78.1	478	1919	0.249	0.273	0.561	0.487
Kia Niro EV	68.0	457	1757	0.260	0.237	0.475	0.499
Tesla Model S	85.0	544	2170	0.251	0.278	0.709	0.393
Ford Mustang Mach-E	75.7	485	2044	0.237	0.593	0.322	1.842
Chevrolet Bolt	60.0	435	1628	0.267	0.306	0.470	0.651
Hyundai Bolt	64.0	452	1760	0.257	0.390	0.332	1.175

J. Maier, *Optimization of Zonal Electric/Electronic Architectures Using a Coupled Modeling Approach of Package and Wiring Harness*, Wissenschaftliche Reihe Fahrzeugtechnik Universität Stuttgart, https://doi.org/10.1007/978-3-658-48188-9

A.2 Electric Component Data in Literature

Table A.2: Electrical loads: ADAS, AD, body and comfort adapted from [151]

Component	$P_{el,max}$ in W	t_{use} in %	$P_{\varnothing}$ in W	n_{veh}	P_{tot} in W	Reference
			ADAS and AD			
Camera	2	100	2	7	14	[195]
ECU	96	100	96	2	192	[195]
DSRC	2	100	2	1	2	[195]
Lidar	60	100	60	2	120	[195]
Radar	8	100	8	8	64	[195, 196]
Sonar/Ultrasound	0.15	100	0.15	2	0.3	[195]
			Body and Comfort I			
Air conditioner front		100	600	1		[197]
Air conditioner rear		100	480	1		[197]
Anti-fog light	35	0	0	3	0	[198]
Blind zone radar	5	100	5	1	5	[198]
Braking light	21	50	10.5	3	31.5	[198]
Cabin lights	20	100	20	1	20	[198]
Central lock	15	0	0	1	0	[198]
Cigarette lighter	100	5	5	1	5	[198]
Door module		100	360	2		[197]
Electric roof	300	5	15	1	15	[198]
Front window heater	1500		120	1		[199]
Front wiper		90	60	1		[200]
Headlamps	60	100	60	2	120	[198]
Horns/Fanfares	100		25	1		[200]
Mirror heater		10	56	2		[201]
Number plate	5	100	5	4	20	[198]
Parking distance control	20	10	2	1	2	[198]

Table A.3: Electrical loads: Body and Comfort, Infotainment, Powertrain and vehicle dynamics adapted from [151]

Component	$P_{el,max}$ in W	t_{use} in %	$P_{\varnothing}$ in W	n_{veh}	P_{tot} in W	Reference
			Body and Comfort II			
Rear window heating			120	1		[202]
Rear wiper		65	30	1		[202]
Reversing light	21	10	2.1	2	4.2	[198]
Seats electronics	300	5	15	2	30	[198]
Turning light	21	20	4.2	6	25.2	[198]
			Infotainment			
Audio system	25	100	25	1	25	[198]
Multimedia screens	30	100	30	2	60	[198]
Navigation	15	100	15	1	15	[198]
			Powertrain and vehicle dynamics			
ABS		100	600	1		[203]
Brake-by-wire	1300		150	1	150	[74]
Cooling pump	1200	100	500	1		[199]
Cooling radiator	500	50	100	1		[202]
Engine ECU		100	700	1	700	[74]
ESP	7800		3000	1		[197]
Steer-by-wire	1400		28	1	28	[74]
Suspension pump	1000	20	200	1	200	[198]
Vehicle control unit	1000		750	1	750	[74]

A.3 Component Data of the autoSHUTTLE

Table A.4: Geometric data components of the autoSHUTTLE (1/2)

Component	x_1 in m	x_2 in m	y_1 in m	y_2 in m	z_1 in m	z_2 in m
Cabin camera	2.4	2.6	0.9	1.1	2.1	2.3
Backup ADAS ECU	3.8	4	0.9	1.1	1.4	1.6
Vehicle control unit	4	4.1	0.9	1.1	1.4	1.6
Suspension ECU	3.9	3.95	0.2	0.4	1.4	1.6
Radar front	0	0.2	0.9	1.1	0.1	0.3
Radar rear	4.8	5	0.9	1.1	0.1	0.3
Rear window heating	3.8	4	1.2	1.3	1.4	1.6
Body module front	1.5	1.7	0.4	0.6	0.5	0.7
Front window heating	1.5	1.7	1.4	1.6	0.5	0.7
DSRC	3.3	3.5	0.4	0.6	0.5	0.7
Body module rear	3.3	3.5	1.4	1.6	0.5	0.7
AC compressor	0.5	0.7	0.9	1.1	0.4	0.6
Cabin lights	3.5	3.7	0.9	1.1	0.9	1.1
Suspension pump	0.3	0.5	0.9	1.1	0.4	0.6
Display	2.4	2.6	1.7	1.9	0.9	1.1
Door module	2.8	3	0	0.2	0.9	1.1
External HMI front	0.2	0.4	0.9	1.1	1.4	1.6
External HMI rear	4.7	4.9	0.9	1.1	1.4	1.6
Headlight left	0	0.2	0.1	0.3	0.9	1.1
Headlight right	0	0.2	1.7	1.9	0.9	1.1
Cooling pump	0.5	0.7	0.9	1.1	0.6	0.8
Multimedia screen side	2.4	2.6	1.7	1.9	0.7	0.9
Multimedia screen front	0.7	0.9	0.9	1.1	0.6	0.8
Seat electronic front	1.1	1.3	0.9	1.1	0.5	0.7
Seat electronic rear	3.7	3.9	0.9	1.1	0.5	0.7
Loudspeaker front	0.1	0.3	0.3	0.5	1.1	1.3
Navigation ECU	0.1	0.3	1.5	1.7	1.1	1.3
Loudspeaker back	3.6	3.8	0.3	0.5	1.1	1.3

Table A.5: Geometric data components of the autoSHUTTLE (2/2)

Component	x_1 in m	x_2 in m	y_1 in m	y_2 in m	z_1 in m	z_2 in m
Backup ADAS ECU	3.6	3.8	1.5	1.7	1.1	1.3
Infotainment ECU left	3.5	3.7	0	0.2	0.9	1.1
Sensor fusion ECU 1	1.4	1.55	0.15	0.25	1.2	1.3
Sensor fusion ECU 2	1.4	1.55	1.75	1.85	1.2	1.3
Sensor fusion ECU 3	3.4	3.55	0.15	0.25	1.2	1.3
Sensor fusion ECU 4	3.4	3.55	1.75	1.85	1.2	1.3
Gateway	3.65	3.75	0.9	1.1	1.4	1.6
Infotainment ECU right	3.5	3.7	1.8	2	0.9	1.1
Audio system	3.5	4	0.3	0.5	0.2	0.3
Steer-by-wire	0.95	1.05	0.15	0.25	0.95	1.05
Brake-by-wire	0.95	1.05	1.75	1.85	0.95	1.05
Engine ECU	3.95	4.05	0.15	0.25	0.95	1.05
5G router	3.95	4.05	1.75	1.85	0.95	1.05
BMS ECU	4.8	5	0.3	0.5	0.3	0.5
ABS	4.4	4.6	0.9	1.1	1.4	1.6
ESP	4.4	4.6	1.5	1.7	1.4	1.6
Drive module front Left	0.1	0.3	0.3	0.5	0.1	0.2
Drive module front Right	0.1	0.3	1.5	1.7	0.1	0.2
Drive module rear Left	3.6	3.8	0.3	0.5	0.1	0.2
Drive module rear Right	3.6	3.8	1.5	1.7	0.1	0.2
Radiator powertrain	0.1	0.3	0.35	0.5	0.2	0.35
Radiator interior	0.1	0.3	1.5	1.65	0.2	0.35
GNSS	2.45	2.5	0.95	1	2.3	2.4
Lidar front left	0	0.05	0	0.05	2.1	2.15
Lidar front right	0	0.05	1.9	1.95	2.1	2.15
Lidar rear left	4.8	4.85	0	0.05	2.1	2.15
Lidar rear right	4.8	4.85	1.9	1.95	2.1	2.15

Table A.6: Electric data components of the autoSHUTTLE (1/2)

Component	V_{el} in V	$P_{\varnothing}$ in W	t_u in %	Safety
Cabin camera	12	0.2	100	0
Backup ADAS ECU	400	500	100	1
Vehicle control unit	400	1000	100	1
Suspension ECU	12	12	30	0
Radar front	12	8	70	1
Radar rear	12	8	30	1
Rear window heating	12	120	100	1
Body module front	400	360	100	1
Front window heating	400	120	100	1
DSRC	400	2	70	1
Body module rear	400	360	100	1
AC compressor	12	600	30	0
Cabin lights	12	20	100	0
Suspension pump	12	1000	20	0
Display	12	30	100	0
Door module	12	360	100	0
External HMI front	12	150	70	0
External HMI rear	12	150	70	0
Headlight left	12	60	100	0
Headlight right	12	60	100	0
Cooling pump	12	500	100	0
Multimedia screen side	12	24	100	0
Multimedia screen front	12	24	100	0
Seat electronic front	12	300	5	0
Seat electronic rear	12	300	5	0
Loudspeaker front	400	70	5	0
Navigation ECU	400	15	100	0
Loudspeaker Back	400	70	50	0

Safety: 1 = critical, 0 = uncritical

Table A.7: Electric data components of the autoSHUTTLE (2/2)

Component	V_{el} in V	$P_{\varnothing}$ in W	t_u in %	Safety
Backup ADAS ECU	400	20	100	0
Infotainment ECU left	12	4	100	0
Sensor Fusion ECU 1	12	96	100	1
Sensor Fusion ECU 2	12	96	100	1
Sensor Fusion ECU 3	12	96	100	1
Sensor Fusion ECU 4	12	96	100	1
Gateway	12	5	100	0
Infotainment ECU right	12	36	100	1
Audio system	12	25	50	0
Steer-by-wire	12	28	100	1
Brake-by-wire	12	1300	100	1
Engine ECU	12	700	100	1
5G router	12	68.4	100	1
BMS ECU	12	24	100	1
ABS	12	600	100	1
ESP	400	3000	50	1
Drive module front Left	400	13000	25	1
Drive module front right	400	13000	25	1
Drive module rear left	400	13000	25	1
Drive module rear right	400	13000	25	1
Radiator powertrain	48	500	50	1
Radiator interior	48	500	50	1
GNSS	12	0.55	100	0
Lidar front left	12	60	100	0
Lidar front right	12	60	100	0
Lidar rear left	12	60	100	0
Lidar rear right	12	60	100	0

Safety: 1 = critical, 0 = uncritical

A.4 Component Data of the Premium SUV

Table A.8: Geometric data components of the premium SUV (1/2)

Component	x_1 in m	x_2 in m	y_1 in m	y_2 in m	z_1 in m	z_2 in m
Radar sensor	0.00	0.10	1.26	1.32	0.2	0.4
Ultrasonic sensor	0.00	0.10	0.93	0.98	0.1	0.3
Headlight left	0.16	0.30	1.64	1.71	0.4	0.6
Headlight right	0.16	0.31	0.19	0.25	0.4	0.6
LiDAR left	0.27	0.39	1.51	1.56	0.4	0.6
LiDAR right	0.26	0.40	0.33	0.39	0.4	0.6
BMS	0.33	0.47	0.93	0.98	0.3	0.5
Motor control unit	0.57	0.72	0.93	0.99	0.3	0.5
AC compressor	0.86	1.02	0.93	0.99	0.2	0.4
Suspension pump	0.93	1.08	1.59	1.65	0.2	0.4
ABS	0.97	1.11	0.33	0.39	0.2	0.4
Steer-by-wire	1.06	1.19	0.93	0.98	0.3	0.5
ESP	1.24	1.38	1.49	1.55	0.2	0.4
Windshield heating	1.39	1.52	1.70	1.76	0.2	0.4
Thermal management engine	1.39	1.52	0.16	0.21	0.1	0.3
Radiator front	1.40	1.54	1.32	1.38	0.2	0.4
Multimedia display front	1.40	1.54	0.56	0.63	0.3	0.5
Gateway	1.49	1.63	0.26	0.32	0.3	0.5
Navigation	1.64	1.78	1.32	1.38	0.3	0.5
Mirror camera left	1.66	1.81	1.72	1.78	0.3	0.5
Mirror camera right	1.66	1.82	0.11	0.17	0.3	0.5
Door module front left	1.86	1.99	1.75	1.81	0.2	0.4
Loudspeaker driver	1.86	1.99	1.36	1.42	1.3	1.5
5G router	1.86	1.99	1.29	1.34	1.3	1.5
Camera	1.87	2.00	0.93	0.99	1.3	1.5

Table A.9: Geometric data components of the premium SUV (2/2)

Component	x_1 in m	x_2 in m	y_1 in m	y_2 in m	z_1 in m	z_2 in m
Door module front right	1.86	1.99	0.10	0.17	0.2	0.4
Seat electronic co-driver	2.03	2.17	0.54	0.60	0.1	0.3
GNSS	2.10	2.22	0.93	0.98	1.3	1.5
Cabin camera	2.32	2.47	1.32	1.38	1.3	1.5
Backup ECU	2.32	2.46	0.94	0.98	0.1	0.3
Interior light	2.32	2.47	0.58	0.64	1.3	1.5
Sensor fusion ECU 1	2.49	2.62	1.32	1.38	0.1	0.3
Sensor fusion ECU 2	2.49	2.62	0.54	0.59	0.1	0.3
Door module back left	2.64	2.78	1.74	1.80	0.3	0.5
Door module back right	2.65	2.78	0.10	0.17	0.3	0.5
Radiator back	2.88	3.02	0.93	0.98	0.2	0.4
Emergency sender	2.98	3.12	0.65	0.72	1.3	1.5
Lighting back	3.64	3.77	0.93	0.99	1.3	1.5
Antenna	3.85	3.99	0.93	0.98	1.3	1.5
Lidar	3.85	3.99	1.51	1.56	1.3	1.5
Loudspeaker back left	4.01	4.14	0.18	0.24	1.3	1.5
Loudspeaker back right	4.10	4.23	1.58	1.63	1.3	1.5
Synthetic motor sound	4.20	4.33	0.18	0.24	0.6	0.8
Comfort ECU	4.38	4.51	1.63	1.69	0.6	0.8
Rear light right	4.68	4.81	1.63	1.69	0.4	0.6
Rear window heater	4.69	4.81	0.93	0.99	0.4	0.6
Rear light left	4.67	4.82	0.23	0.29	0.4	0.6
Axle drive front	0.20	0.50	0.92	1.00	0.2	1
Axle drive back	4.50	4.80	0.92	1.00	0.2	1
PTC 1 engine	0.30	0.40	0.50	0.60	0.3	0.5
PTC 2 cabine	0.80	0.90	0.50	0.60	0.3	0.5

Table A.10: Electric data components of the premium SUV (1/2)

Component	V_{el} in V	$P_{\varnothing}$ in W	t_u in %	Safety
Radar sensor	12	8	1	1
Ultrasonic sensor	12	0.15	1	1
Headlight left	12	60	1	0
Headlight right	12	60	1	0
Lidar left	12	60	1	1
Lidar right	12	60	1	1
BMS	12	24	1	1
Motor control unit	400	1000	1	1
AC compressor	12	600	0.3	0
Suspension pump	12	1000	0.3	0
ABS	12	600	1	1
Steer-by-wire	12	28	1	0
ESP	400	3000	1	1
Windshield heating	12	120	0.1	0
Thermal management engine	12	120	1	0
Radiator front	400	500	1	0
Multimedia display front	12	24	1	0
Gateway	12	5	1	0
Navigation	12	600	1	0
Mirror camera left	12	2	1	0
Mirror camera right	12	2	1	0
Door module front left	12	360	1	0
Loudspeaker driver	12	70	0.8	0
5G router	12	12	1	0
Camera	12	2	1	0

Safety: 1 = critical, 0 = uncritical

Table A.11: Electric data components of the premium SUV (2/2)

Component	V_{el} in V	$P_{\varnothing}$ in W	t_u in %	Safety
Door module front right	12	360	1	0
Seat electronic co-driver	12	300	0.5	0
GNSS	12	0.55	1	0
Cabin camera	12	0.2	1	0
Backup ECU	400	500	1	1
Interior light	12	20	0.5	0
Sensor fusion ECU 1	12	96	1	1
Sensor fusion ECU 2	12	96	1	1
Door module back left	12	360	1	0
Door module back right	12	360	1	0
Radiator back	400	500	0.3	0
Emergency sender	12	1	0.1	1
Lighting back	12	20	0.1	0
Antenna	12	10	1	0
Lidar	12	60	1	1
Loudspeaker back left	12	70	0.5	0
Loudspeaker back right	12	70	0.5	0
Synthetic motor sound	12	70	1	0
Comfort ECU	400	360	1	0
Rear light right	12	21	0.3	0
Rear window heater	12	120	0.1	0
Rear light left	12	21	0.3	0
Axle drive front	400	70000	0.5	1
Axle drive back	400	95000	0.5	1
PTC 1 engine	400	1000	0.3	0
PTC 2 cabine	400	1000	0.5	0

Safety: 1 = critical, 0 = uncritical

B. Algorithm Parameters

This section presents parameter and corresponding data of the algorithms used in this thesis.

B.1 Genetic Algorithm - Aggregation Tree

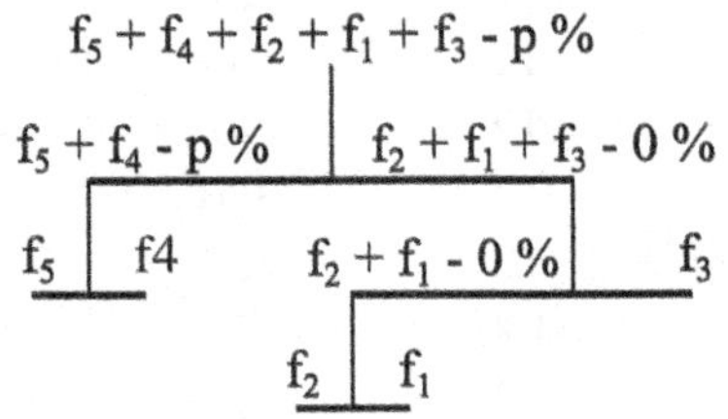

Figure B.1: Aggregation Tree for f_1 to f_5

B.2 k-Means Clustering - Initial Cluster Positions Based on Heuristic

Figure B.2 shows the initial cluster position for the k-means clustering based on vehicles boundaries and preferences. The positions are selected using three different strategies. Set A strives for an equal spatial distribution, Set B for an equal distribution by means of the angle in the x-y plane and C comprises various additional promising variants. Set X is the combination of the best performing positions for each zone number based on the evaluation by Manhattan distance.

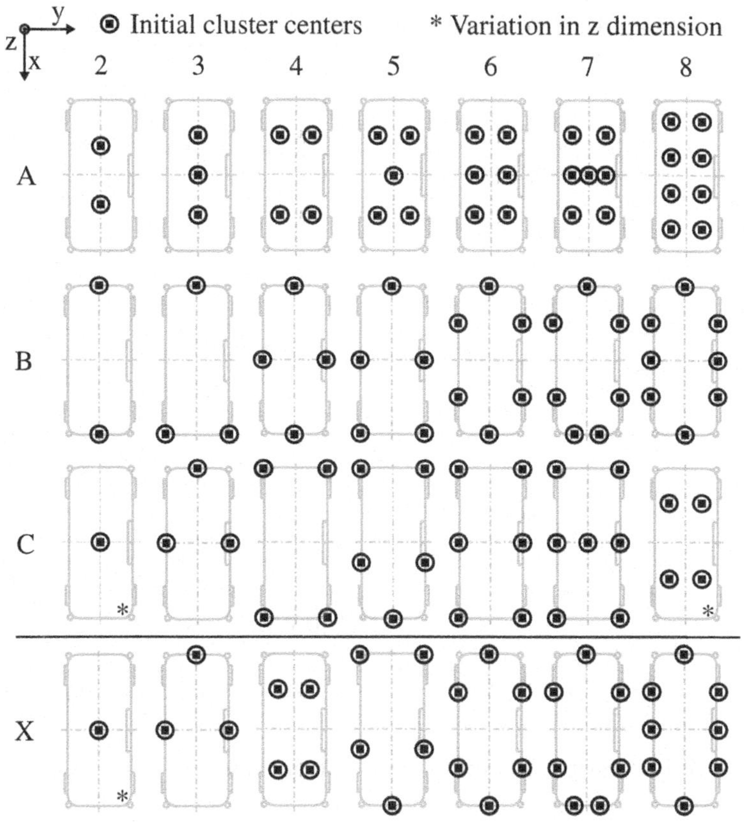

Figure B.2: Initial cluster positions based on vehicles boundaries and preferences for zone numbers of two to eight

B.3 Low Voltage Battery Regression Data

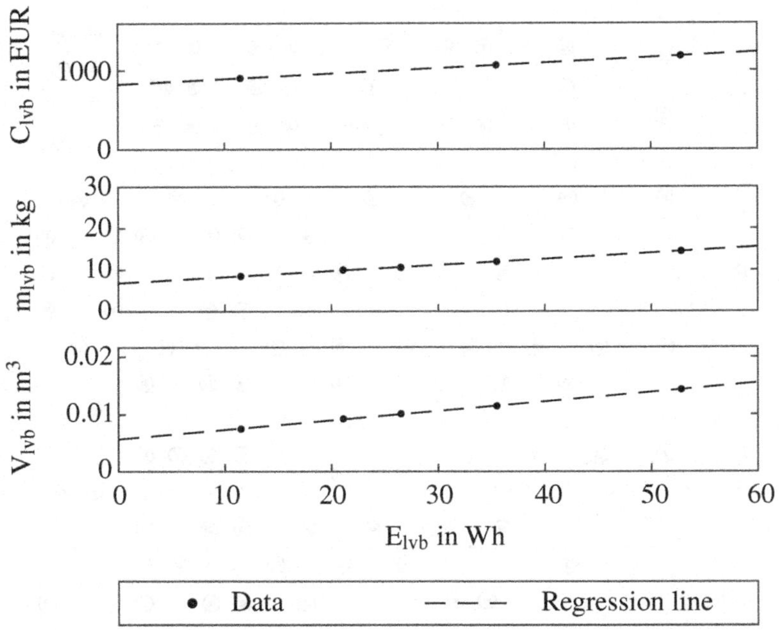

Figure B.3: Regression data for the low voltage battery

The data points are extracted from available 12 V lead acid batteries. The regression lines are determined using the least square method. The equations of the lines and their coefficients of determination are listed below:

- Cost: $C_{sc} = 0.8902\frac{EUR}{Ah}E_{sc} + 47.595EUR$, R^2=0.2077
- Mass: $m_{sc} = 0.2701\frac{kg}{Ah}E_{sc}$, R^2=0.8173
- Volume: $V_{sc} = 0.1067\frac{m^3}{Ah}10^{-3}E_{sc} + 1.6003m^3 10^{-3}$, R^2=0.9424

B.4 Supercapacitor Regression Data

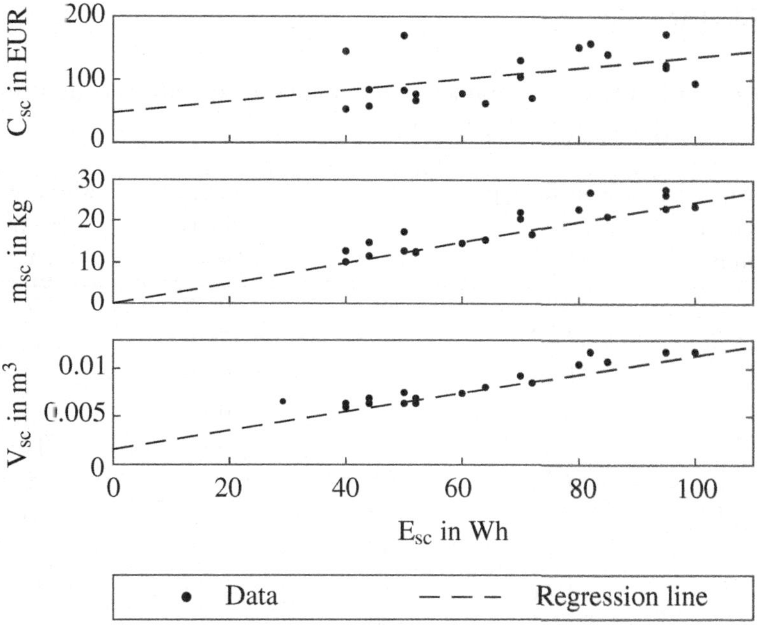

Figure B.4: Regression data for the supercapacitor

The data points are extracted from available 48V SCs and the regression lines are calculated using the least square method. The resulting equations of the lines and their coefficients of determination are listed below:

- Cost: $C_{sc} = 6.7953\frac{EUR}{Ah}E_{sc} + 824.55EUR$, R^2=1.0
- Mass: $m_{sc} = 0.1447\frac{kg}{Ah}E_{sc} + 6.854kg$, R^2=0.9992
- Volume: $V_{sc} = 0.1658\frac{m^3}{Ah}10^{-3}E_{sc} + 5.6716m^3 10^{-3}$, R^2=0.9991

C. Model Data

C.1 Electric Cable

Table C.1: Electric cable parameters for common wire cross-sections in vehicles according to [28]

Cross section [mm^2]	Resistance p. meter [$\frac{m\cdot\Omega}{m}$]	Mass p. meter [$\frac{kg}{m}$]	Cost p. meter [$\frac{EUR}{m}$]	Maximum current [A]
0.5	37.1	9	0.45	6
0.75	24.7	12	0.6	8
1	18.5	15	0.75	12
1.25	14.9	15	0.75	14
1.5	12.7	20	1	16
2	9.42	26	1.3	20
2.5	7.6	32	1.6	24
3	6.15	38	1.9	30
4	4.71	49	2.45	40
5	3.94	60	3	45
6	3.14	69	3.45	50
8	2.38	90	4.5	56
10	1.82	113	5.65	80
12	1.52	144	7.2	90
16	1.16	181	9.05	100
20	0.955	221	11.05	110
30	0.647	325	16.25	130
35	0.527	361	18.05	140
40	0.473	438	21.9	150
50	0.368	521	26.05	160
60	0.315	644	32.2	180
70	0.259	716	35.8	200
95	0.196	918	45.9	260
120	0.153	1220	61	300

C.2 Data Cable

Table C.2: Data cable parameters, assumptions based on [204, 205]

Type	Mass p. meter $[\frac{kg}{m}]$	Cost p. meter $[\frac{EUR}{m}]$
10BASE-T1S	0.03	0.2
1000BASE-T1	0.06	0.4

C.3 Wiring Harness

Table C.3: Wiring harness parameters

Parameter	Value	Unit
Breakout cost	3	$\frac{EUR}{1}$
Angle cost	1	$\frac{EUR}{1}$

C.4 Geometric Model

Table C.4: Parameters for the geometric modeling

Description	Symbol	Value	Reference
Packaging efficiency mass cylindrical, pouch, prismatic	$\eta_{bat,m}$	1.14, 1.23, 1.12	[80]
Battery auxiliaries cost	$C_{bat,aux}$	200 EUR	[135]

C.5 NSGA-II Settings

The settings are based on the default values implemented by Mathworks and developed by Deep et. al [206, 207].

- **Population Size:** 50

- **Parent selection function:** Tournament selection based on a random selection of four elements.

- **Crossover rate:** 0.8

- **Crossover function:** The crossover function consists of two functions. Every time a children is generated one of them is selected randomly.

$$G_c = \begin{cases} G_{p,1} + g_{cf}\left|G_{p,1} - G_{p,2}\right| & \text{and} \\ G_{p,2} + g_{cf}\left|G_{p,1} - G_{p,2}\right| & , \end{cases} \qquad \text{Eq. C.1}$$

  wherein G_c is the produced children, $G_{p,1}$ and $G_{p,2}$ are the parents as well as g_{cf} is a random number generated from a Laplace distribution.

- **Mutation rate:** Adapted by mutation function, decreases by rising iteration number.

- **Mutation function:** The mutation function creates a child based on the mutation of a parent. The following equation mutates every component i separately according to:

$$G_{c,i} = \begin{cases} \Delta_{lb} < g_{mf,1} & G_{p,i} - g_{mf,2}(G_{p,i} - lb(i)) \quad \text{and} \\ \Delta_{lb} \geq g_{mf,1} & G_{p,i} - g_{mf,2}(ub(i) - G_{p,i}) \quad . \end{cases} \qquad \text{Eq. C.2}$$

  Therein Δ_{lb} is the scaled distance between the ith element of the parent and the lower boundary, $g_{mf,1}$ as well as $g_{mf,2}$ are random variables from an uniform and a power distribution, lb is the lower and ub the upper boundary.

C.6 Consistency with Initial Clustering

Figure C.1 visualizes the normalized cluster consistency when random objects are removed from the data set. The total number of objects in the data set is 55 and the maximum number of removed objects is 39, which is a share of 70 %. The cluster consistency decreases first exponential and then linear, while the range of the exponential decrease expands for higher zone numbers. This results in lower values for higher zone numbers, however the seven zone architecture is an exception. The values are determined based on the average results of 1000 iterations and the removed objects were selected randomly. The Flexibility is determined by the average value of the results from one to 15 removed objects.

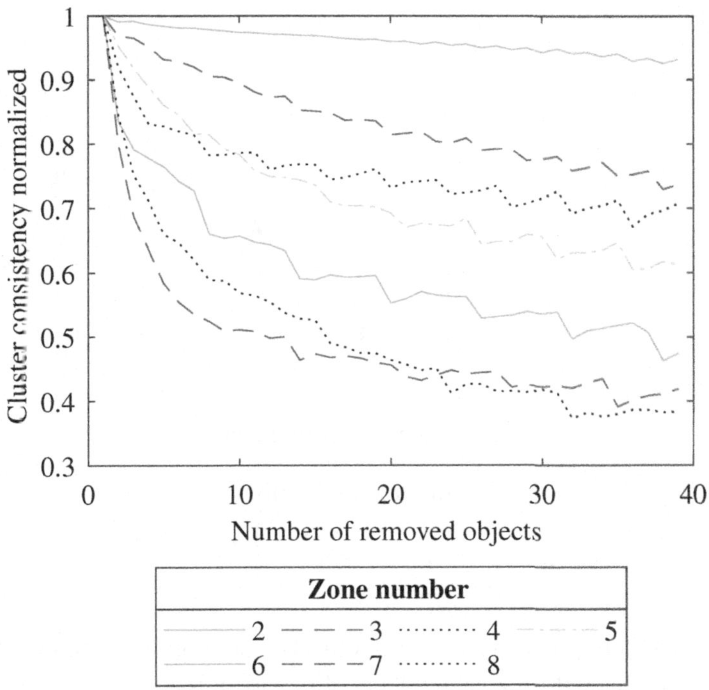

Figure C.1: Cluster consistency in dependency of the number of removed data objects for zone numbers of two to eight

C.7 Minimum Distance for Cluster Resorting

Figure C.2 shows the minimum distance a object needs to be shifted to be assigned to another cluster. The objects was selected by the the minimum distance to any other cluster centroid and the step size was 0.01 m in each coordinate direction. While the values were not used to determine the Flexibility, they indicate that the necessary distance is significant higher for zone numbers of six to eight. For the average of all objects this was expected, but as this simulation only considers the critical data points the results are remarkable.

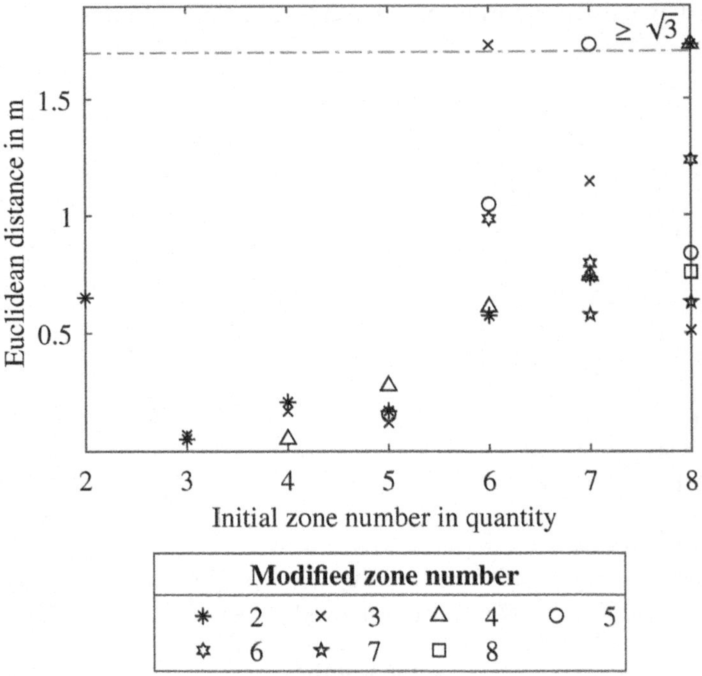

Figure C.2: Minimum displacement of objects resulting in a different cluster alignment for zone numbers of two to eight

D. Model Feasibility

Models approximate real world systems and simplify them to reduce calculation and modeling effort. These simplification can result in inaccurate or implausible results. This chapter discusses two critical aspects of the model. The first is the bending radius, which is critical by means of the geometric modeling of wiring harnesses. The second is the voltage drop in a wire, which limits the physical characteristics, especially cross-section and length.

D.1 Bending Radius

The bending radius is not considered in the wiring harness design respectively the cable routing, but it is inevitable to ensure compatibility to industrial standards. Therefore this thesis follows the recommendations of the German Electrical and Electronic Manufacturers‘ Association on the bending radius of cables in vehicles [208]. They introduce a set of rule for both, static and dynamic applications. The limits of the dynamic case are selected for the verification of the geometric grid model, as it has tougher threshold values than the static case. The following equations show a worst case inspection for the highest cable diameter of the data set and the dynamic case.

Largest cable cross-section:

$$A_{cab,max} = 95mm^2 \qquad \text{Eq. D.1}$$

Largest cable diameter:

$$D_{cab,max} = 2\sqrt{\frac{A}{\pi}} = 16.7mm \qquad \text{Eq. D.2}$$

Minimum bending radius for dynamic load:

$$r_{b,dyn} = 6D = 0.1002m \qquad \text{Eq. D.3}$$

Grid approximation:

$$d_{grid} = 0.1m \qquad \text{Eq. D.4}$$

Deviation of the approximation:

$$\Delta_b = r_{b,dyn} - d_{grid} = 0.0002m \qquad \text{Eq. D.5}$$

D.2 Voltage Drop

As the electric load is changing dependent on the active functions, the power supply can not keep the voltage constant. Therefore vehicle components are developed to operate reliable in a specific voltage range. Cables, fuses and other connecting elements behave, from an electrical point of view, like resistors causing a voltage drop. To ensure unrestricted operation, the voltage drop must be relatively low compared to the operation voltage. Consequently the voltage drop was reviewed for two worst case scenarios, one for HV and another for LV.

The most powerful HV component are the drives. Thus they were selected, and adapted to form a worst-case scenario. As many EVs have two electric drives, the same is assumed here, therefor the initial current of a single drive unit is doubled. The length was derived from the front left wheel to the end of the drivers cabin, on the longitudinal axis near the vehicle ground. The following equations show a simplified calculation of the voltage drop in the HVs power supply.

$$R_{el,cab} = \frac{l_{cab}\rho_{Cu}}{A_{cab}} = \frac{4.2m \cdot 0.01786\frac{\Omega mm^2}{m}}{10mm^2} = 0.0075012V \quad \text{and} \qquad \text{Eq. D.6}$$

$$V_{el,cab} = R_{cab}I_{max} = 0.0075012\Omega \cdot 72A = 0.54V \quad , \qquad \text{Eq. D.7}$$

wherein ρ_{Cu} is the specific resistance of copper, l_{cab} and A_{cab} are length and cross-section, $R_{el,cab}$ and $V_{el,cab}$ are resistance and voltage drop of the cable and $I_{HV,max}$ is the maximum current of all HV components. The voltage drop of 0.54 V corresponds to an relative drop of 0.135 % and thus is not critical.

The most powerful LV component is the anti-lock braking system. For a worst case scenario the cable length is set as the same as for the HV level, albeit it should be shorter as the central anti-lock braking system typically is positioned nearer to the longitudinal axis. The voltage drop for the low voltage side is determined using the following equations:

$$R_{cab,LV} = \frac{l_{cab,LV}\rho_{Cu}}{A_{cab,HV}} = \frac{4.2m \cdot 0.01786\frac{\Omega mm^2}{m}}{95mm^2} = 0.0007896V \quad \text{and} \qquad \text{Eq. D.8}$$

$$U_{cab,LV} = R_{cab,LV} I_{LV,max} = 0.0007896\Omega \cdot 240A = 0.1895V \quad , \qquad \text{Eq. D.9}$$

wherein $l_{cab,LV}$ and $A_{cab,LV}$ are length and cross-section, $R_{cab,LV}$ and $U_{cab,LV}$ are resistance and voltage drop of the cable and $I_{HV,max}$ is the maximum current of all LV components. The voltage drop of 0.1895 V corresponds to an relative drop of 1.578 % and thus is also not critical.

E. Additional Evaluation Data

In this section additional results are listed, in the following order:

Table E.1: Figure list overview

Vehicle type	Objective	Modification	Figure name
autoSHUTTLE	Core properties	Powertrain	Figure E.1
autoSHUTTLE	Structure properties	Powertrain	Figure E.2
autoSHUTTLE	Core properties	Battery	Figure E.3
autoSHUTTLE	Structure properties	Battery	Figure E.4
autoSHUTTLE	Design structure matrix	various	Figure E.5

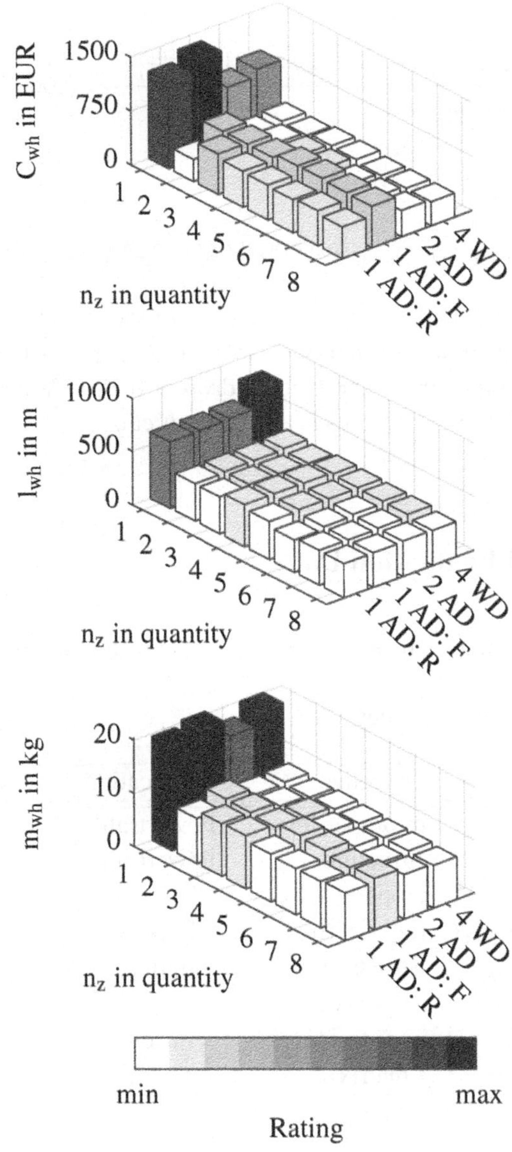

Figure E.1: Wiring harness core properties of different powertrain concepts for the autoSHUTTLE

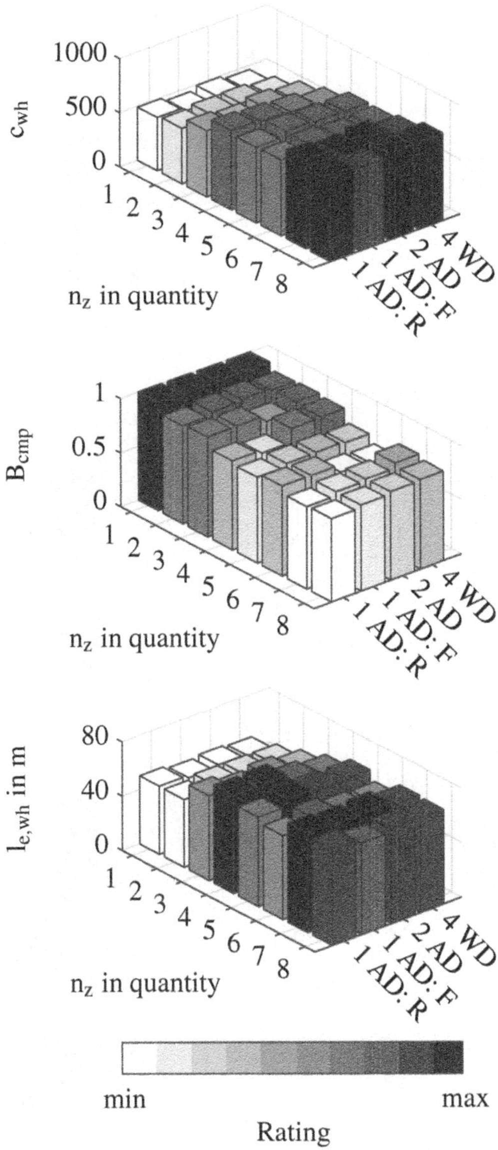

Figure E.2: Wiring harness structural properties of different powertrain concepts for the autoSHUTTLE

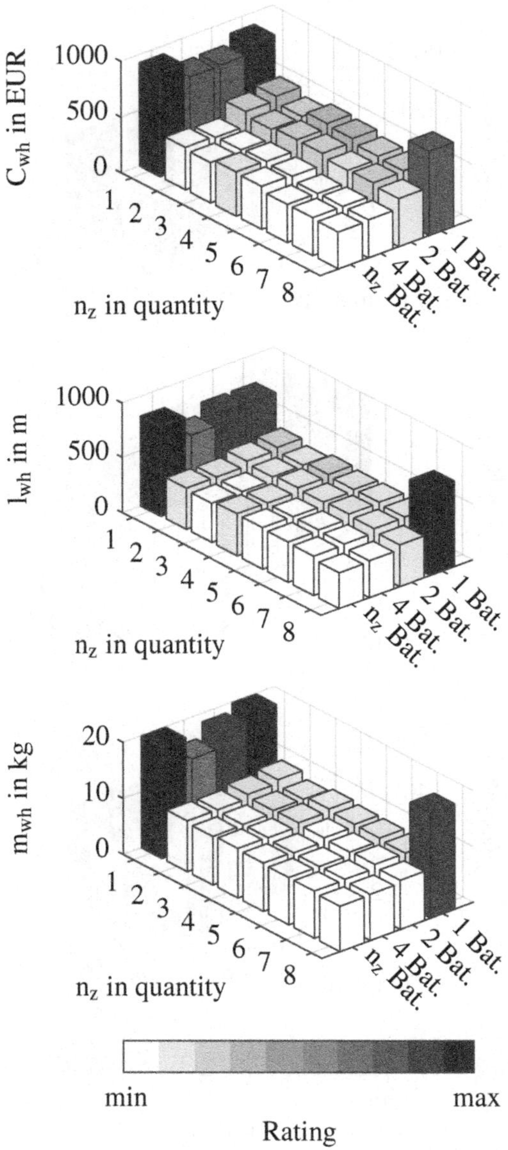

Figure E.3: Wiring harness core properties of different door concepts for the autoSHUTTLE

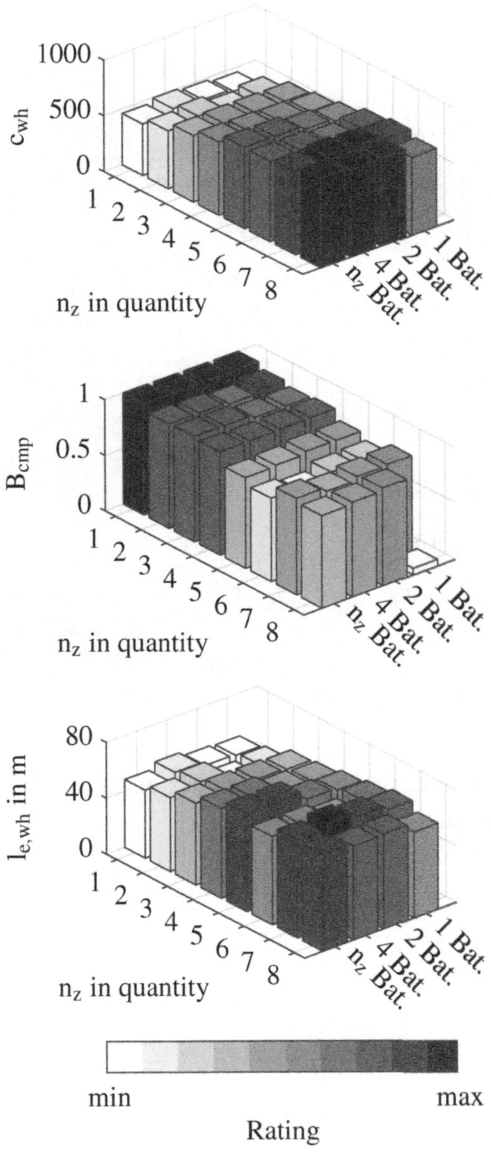

Figure E.4: Wiring harness structural properties of different door concepts for the autoSHUTTLE

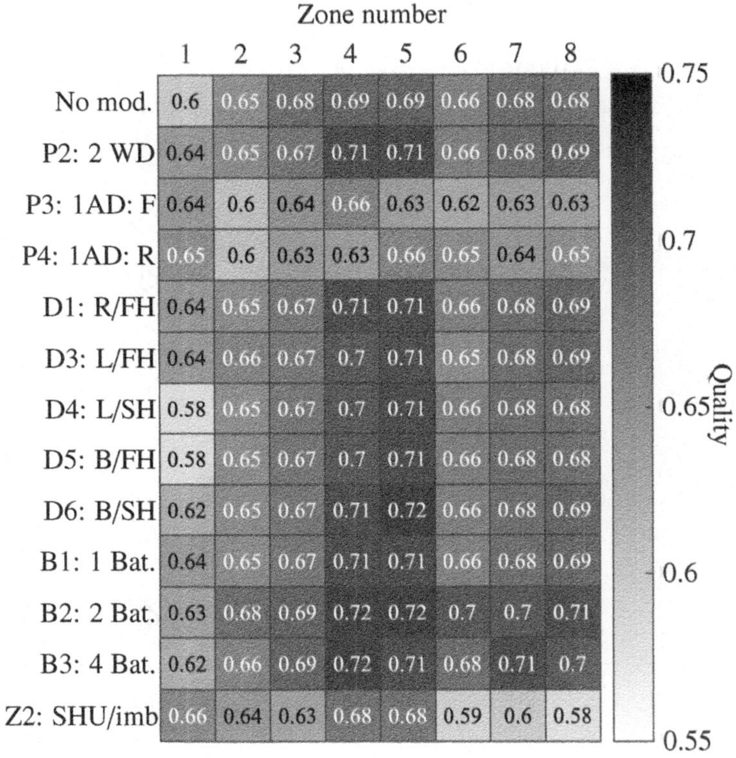

Figure E.5: Reduced design structure matrix of the autoSHUTTLE with equal metric weighting

The manufacturer's authorised representative in the EU is Springer Nature Customer Service Centre GmbH, Europaplatz 3, 69115 Heidelberg, Germany. If you have any concerns regarding our products, please contact ProductSafety@springernature.com

Printed and bound by CPI Group (UK) Ltd, Croydon, CR0 4YY
30/09/2026
02225566-0002